# Bioresorbable Materials and Their Application in Electronics

Bioresorbable electronics that can dissolve away in aqueous environment and generate biological safe products offers revolutionary solutions to replace current built-to-last electronics predominantly used in implanted devices and electronic gadgets. Its contribution involves reducing risk of surgical complications by minimizing number of surgery and preventing production of electronic waste by allowing rapid device recycling. This Element presents bioresorbable materials (e.g. metals, polymers, inorganic compounds, and semiconductors) that have been used to construct electronic devices and analyzes their unique dissolution behaviors and biological effects. These materials are combined to yield representative devices including passive and active components and functional systems.

# BIORESORBABLE MATERIALS AND THEIR APPLICATION IN ELECTRONICS

Xian Huang
*Tianjin University*

## ELEMENTS OF FLEXIBLE AND LARGE-AREA ELECTRONICS

Ravinder Dahiya

and

Luigi Occhipinti

CAMBRIDGE
UNIVERSITY PRESS

Cambridge Elements

# CAMBRIDGE
## UNIVERSITY PRESS

University Printing House, Cambridge CB2 8BS, United Kingdom

One Liberty Plaza, 20th Floor, New York, NY 10006, USA

477 Williamstown Road, Port Melbourne, VIC 3207, Australia

314–321, 3rd Floor, Plot 3, Splendor Forum, Jasola District Centre,
New Delhi – 110025, India

79 Anson Road, #06–04/06, Singapore 079906

Cambridge University Press is part of the University of Cambridge.

It furthers the University's mission by disseminating knowledge in the pursuit of
education, learning, and research at the highest international levels of excellence.

www.cambridge.org
Information on this title: www.cambridge.org/9781108406239
DOI: 10.1017/9781108290685

First published 2018

A catalogue record for this publication is available from the British Library.

ISBN 978-1-108-40623-9 Paperback
ISSN 2398-4015

Cambridge Elements ≡

# Bioresorbable Materials and Their Application in Electronics

Xian Huang

*Biomedical Engineering, School of Precision Instrument and Opto-electronic Engineering, Tianjin University, Tianjin, China, 300072*

**Abstract:** *Bioresorbable electronics that can dissolve away in aqueous environments and generate biologically safe products offers revolutionary solutions for replacing current built-to-last electronics predominantly used in implanted devices and electronic gadgets. Its contribution involves reducing the risk of surgical complications by minimizing the number of surgeries required and preventing the production of electronic waste by allowing rapid device recycling. This text presents bioresorbable materials (e.g. metals, polymers, inorganic compounds, and semiconductors) that have been used to construct electronic devices and analyzes their unique dissolution behaviors and biological effects. These materials are combined to yield representative devices including passive and active components and functional systems.*

**Key words:** *Bioresorbable materials, bioresorbable electronics, transient electronics, biomedical implants, dissolution.*

---

## 1 Introduction to Bioresorbable Devices

The dissolution of materials in aqueous media is a common phenomenon that has inspired numerous applications in daily life and scientific research. Some materials can dissolve in biofluids and generate products that are biologically safe, and, thus, are termed as bioresorbable materials. Modern applications of bioresorbable materials include biomedical implants (e.g. stent[1, 2], fixation[3, 4], scaffold[5, 6], bone graft[7, 8], and suture[9–11]) (Figure 1), food/medicine

1

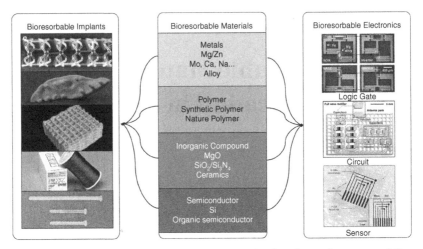

**Figure 1.** Lists of bioresorbable materials that have been used for biomedical implants and bioresorbable electronics.

packages[12, 13], and additives to cosmetic and pharmaceutical products[14]. These applications have inspired the use of bioresorbable materials in the recent development of bioresorbable electronics (or transient electronics)[15] (Figure 1) that can disappear into the surrounding environment via controlled physical or chemical changes with defined rates (from minutes to weeks) (Figure 2).

Emerging bioresorbable electronics undergoes several stages, and benefits from the development of bioresorbable materials that were originally used in biomedical implants. Traditional biomedical implants are realized by nonabsorbable materials, including metals and polymers such as stainless steel[16-18], cobalt alloys[19-21], titanium and its alloys[22-24], and carbon fiber composites[25-27]. These materials offer high corrosion resistance in the implantation environment, but may release toxic chemicals[28] such as Fe(III), Ti(IV), Cr(VI), and Co(II) during degradation processes[29], causing local inflammation[30], osteolysis[31], hypersensitivity[32], and neuropathy[33]. As a result, bioresorbable metals, polymers, and inorganic compounds are introduced as alternative materials for biomedical implants. The addition of bioresorbable semiconductors further enriches

(a)                                             (b)

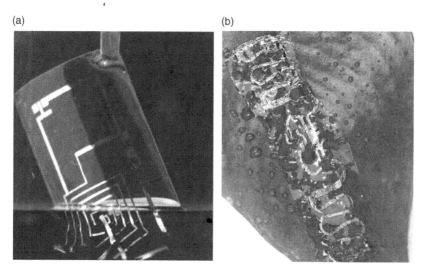

**Figure 2.** Examples of bioresorbable (transient) electronics devices. (a) A transient rectification circuit on a silk substrate that is dissolving in water (photo: John A. Rogers). (b) A screen-printed bioresorbable antenna based on zinc microparticles on sodium carboxymethyl cellulose. The antenna can be dissolved away by rinsing with water
(photo: Xian Huang).

material selection, allowing the appearance of bioresorbable electronics that contains both passive and active electronic components. Currently, more materials are being added to the list of materials that can be used to make bioresorbable electronic devices.

Some pioneering work in developing bioresorbable electronics involves the development of partly resorbable electronic devices[34–36] and the investigation of various types of bioresorbable materials that are suitable for constructing transient electronic devices[37–40]. The collective effort of the research mentioned above has yielded the development of fully bioresorbable electronic devices, which have been demonstrated both as basic components[15, 41] including batteries, sensors, passive components (capacitor, inductor, antenna, and resistor)[42, 43], and active components (diode and MOSFET)[43, 44], and as simply functional systems[45, 46].

Bioresorbable electronics can potentially make incredible strides in the next few years, and hold the promise of making revolutionary advances in health care and environmental protection, resulting in enormous social and economic impacts. Currently, it is estimated that there are annually 16 million surgical procedures performed in acute care hospitals in the United States[47]. These procedures account for ~0.5 million infections annually and nearly four million extra hospital days, as well as $2 billion in increased health care costs[48]. Bioresorbable electronics can be used to conduct internal post-surgical monitoring of organ, tissue, implant, and wound health without the need for surgery to remove them, thus reducing the risk of infection and complications during and after the surgical processes. With regard to environmental protection, increasing demands for electronic gadgets have accelerated the rate of obsolescence, leading to volumes of expended electronic waste (25 million tons $yr^{-1}$)[49], the majority (82%) of which ends up in landfills[50]. The electronic waste in landfills undergoes extremely slow degradation processes, releasing toxic materials such as lead[51], mercury[52], and brominated flame retardants[53] that pollute ground water and soil[54, 55]. Bioresorbable circuits can be used to replace current built-to-last circuits, and can degrade without releasing harmful toxins, allowing rapid recycling of electronic components on the circuits and further reduction in environmental pollution due to unsustainable recycling procedures[56, 57]. The scope of this text mainly involves presenting constituent materials for bioresorbable biomedical implants in general, with the emphasis on materials that have been used in making bioresorbable electronics, including metals, polymers, inorganic compounds, and semiconductors (Figure 1). The biological effects and dissolution mechanisms of these materials are also presented, along with representative devices that demonstrate recent achievements in bioresorbable electronics.

## 2  *Materials used in Bioresorbable Devices*

Bioresorbable devices can be made of various bioresorbable materials including metals, nonmetals, and semiconductors. These materials possess varied water dissolution rates and biological effects, and are used with doses designed to allow tolerable amounts of the dissolution products of these materials in the body[58-60]. For bioresorbable electronics, most of the materials are used as thin films or membranes with thickness ranging from tens of nanometers to micrometers to match the life span of specific devices and to facilitate dissolution.

## 2.1  Metals

Bioresorbable metals typically include magnesium (Mg), zinc (Zn), molybdenum (Mo), calcium (Ca), and sodium (Na), all of which can react with water molecules through hydrolysis to generate products such as $Mg(OH)_2$, $Zn(OH)_2$, $H_2MoO_4$, $Ca(OH)_2$, and NaOH. Some of the dissolution products corrode slowly in aqueous conditions, but react with chloride ions in physiological conditions to produce metal chlorides and hydrogen gas. The reported daily tolerable rate of hydrogen gas release in human bodies is 10 $\mu$L $cm^{-2}$[61], which can be used to determine the proper dose and dissolution rate of the bioresorbable metals. The bioresorbable metals are known to be relatively non-cytotoxic in limited quantities, and present naturally in human bodies as vital elements for cell functions.

### 2.1.1  Mg

Magnesium (Mg) is the second most abundant element in cellular systems, and is involved in all metabolic pathways. It stabilizes DNA and chromatin structures, and acts as an essential cofactor in almost all enzymatic systems related to DNA processing[62]. In addition, $Mg^{2+}$ acts in the formation of biological apatite, which determines the extent of bone fragility, bone healing, and regeneration. Mg offers high specific strength and an elastic modulus that is similar to human bone. As a result, Mg is traditionally used as bone grafts[63, 64] and fixation[65, 66]. In addition, its high biocompatibility leads to its intensive use for intervascular stents for the treatment of arterial disease[67-69] (Figure 3a). Mg is the major composition material in bioresorbable electronics, and has been frequently used to make conductive traces, electrodes, and

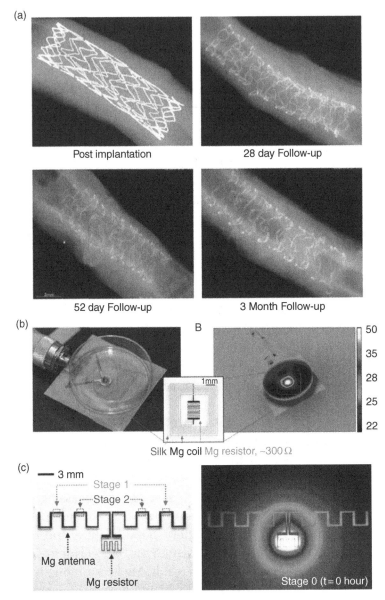

**Figure 3.** Examples of Mg-based biomedical devices that can be used as (a) bioresorbable stents (reprinted with permission from Ref.[68], copyright 2008 Elsevier Inc.), (b) wirelessly actuated heaters for wound care (reprinted with permission from Ref.[70],

passive components including capacitors, antennas, resistors, and coils[15, 42, 44, 70] (Figures 3b and 3c).

Mg possesses rapid *in vivo* corrosion and dissolution rates. When exposed to water, unprotected Mg will produce magnesium hydroxide $(Mg(OH)_2)$[71], which dissolves slowly in water and forms a surface coating on the Mg, causing a reduced corrosion rate of Mg. However, $Mg(OH)_2$ can react with $Cl^-$ in biological environments to form highly soluble Mg chloride $(MgCl_2)$ and hydrogen $(H_2)$ gas. The following reactions can be used to summarize the corrosion reactions of Mg: $Mg + 2H_2O \rightarrow Mg(OH)_2 + H_2$; (1) $Mg + 2Cl^- \rightarrow MgCl_2$; (2) $Mg(OH)_2 + 2Cl^- \rightarrow MgCl_2$[72, 73]. Song and Song measured the corrosion rate of Mg in stimulated body fluid (SBF) using a weight loss measurement approach, and obtained a daily dissolution rate of 19–44 mg cm$^{-2}$, which could be reduced by half when adding phosphate into the SBF[74]. At the reported dissolution rate, the surface areas of the Mg implants are limited to 9–21 cm$^2$, considering that the daily exposure limit of Mg for an average 60-kg adult is about 350–400 mg[75]. Additional influences from the environment can alter the dissolution behavior of Mg by means of different physicochemical parameters (e.g. pH, ion concentrations, oxygen). In addition, dissolution of Mg is connected with the immunological response from the localized tissues due to direct and intimate contact. For example, implanted Mg wires exhibited high corrosion when placed in direct contact with the arterial wall but were not corroded after exposure to blood in the arterial lumen for 3 weeks[76].

The rapid corrosion rate of Mg in the electrolytic physiological environment is one of the greatest limitations for its use in

---

**Caption for Figure 3.** (cont.)

copyright 2011 National Academy of Sciences), and (c) frequency tunable antenna, whose frequency can be changed by partially dissolving antenna structures
(reprinted with permission from Ref.[42], copyright 2015 John Wiley and Sons).

long-term implantable applications. However, this issue can be handled by introducing bioresorbable surface coatings. Typical surface coating materials for Mg include calcium phosphate[77, 78], polycaprolactone[79], silicon carbide (SiC)[80], and silicon dioxide $(SiO_2)$[81], which are deposited onto Mg implants through soaking[77, 78], electrodeposition[82], spraying[79], chemical vapor deposition[80], and physical vapor deposition[83, 84]. By adjusting the coverage and thickness of the coating, the preferential degradation area and rate can be programmed to allow function transformation[42] during different dissolution periods (Figure 3c) and varied working times to adapt to specific implantation needs[85].

The dissolution rate and mechanical strength of Mg can also be tuned by adding bioresorbable alloying elements. An appropriate alloying composition can improve the corrosion resistance, mechanical properties, and ease of manufacture of Mg-based materials. Researchers have developed binary alloys containing two bioresorbable metals such as Mg–Ca[86], Mg–Zn[87], Mg–Nd[88], and Mg–Gd[89], all of which offer moderate resistance to corrosion (2 mm $yr^{-1}$) and mechanical yield strength (<150 MPa). Improved corrosion resistance and mechanical strength can be achieved by introducing multi-element Mg-based alloys. A notable example is the Mg–Zn–Ca alloy, within which Ca improves the corrosion resistance of the alloy and Zn increases the strength of the alloy[90]. A Young's modulus of ~40 GPa and glass transition temperatures in the range 119–129 °C can be achieved by properly selecting the ratio of individual elements in the alloy[91].

### 2.1.2  Zn

Zinc (Zn) is another essential element for human beings, and is a promising alternative to Mg owing to its slower dissolution rate in biological environments[92]. Zn participates in several basic biological functions in human bodies, such as cell growth[93], protein synthesis[94], and DNA replication[95]. Deficiency of Zn may result in skeletal growth retardation and alteration in bone tissue calcification[96, 97]. *In vivo* experiments involving implanting zinc wires in the abdominal aortas of rats indicated low cellular density and lack of smooth-muscle cells adjacent to the implant interface, suggesting an anti-proliferation effect and the prevention of restenosis from zinc implants[98].

Zn offers much slower corrosion rates in biological environments compared with Mg. According to the standard electrode potential, i. e. Mg (-2.37 V)[99] <Zn (-0.763 V)[100]<Fe (0.440 V)[101], Zn has a corrosion rate faster than Fe, but slower than Mg. Four Zn wire samples were implanted in the abdominal aorta of a Sprague-Dawley rat for 1.5, 3, 4.5, or 6 months, respectively. The wires showed uniform corrosion behavior for the first 3 months, and severe, localized corrosion after 4.5 months (Figure 4a). Dissolution rates quantified from measuring the reduction of cross-sectional areas have shown that the average annual corrosion rate of Zn wires varied from ~12 $\mu m\ yr^{-1}$ in the first month to ~50 $\mu m\ yr^{-1}$ after implantation for 6 months[102] (Figure 4b). Another study that involved immersing a Zn coupon 6 mm in diameter and 2 mm in thickness in SBF at pH=7 (37 °C) for 14 days further confirms that the annual corrosion rate of Zn is ~50 $\mu m\ yr^{-1}$[103]. Such corrosion rates are much slower than those of Mg, and result in an average daily release of ~0.1 mg $cm^{-2}$ of zinc implants, limiting the surface area of Zn implants to be less than 80–250 $cm^2$, based on the recommended dietary intake of 8–25 mg $day^{-1}$[104]. Compared to SBF, Zn exhibits less noble dissolution behavior and reduced dissolution rate after immersion in whole blood for 72 h. The reduced dissolution rate of Zn in whole blood is considered to result from the passivation effect from the biomolecules and inorganic dissolution products[104]. Hydrolysis of Zn generates $Zn(OH)_2$, which is again slightly soluble in water, but can dissolve readily in biological solutions that contain $Cl^-$ ions.

Zn has been used as a single construction material for stents[98, 102, 105, 106] and scaffolds[107] (Figure 4c), or used as an alloying material when combined with Mg[87, 108] or Ca[109]. Applications of Zn in bioresorbable electronics range from electrodes[38, 110] for batteries and sensors to micro/nano fillers[111] for conductive pastes (Figure 4d).

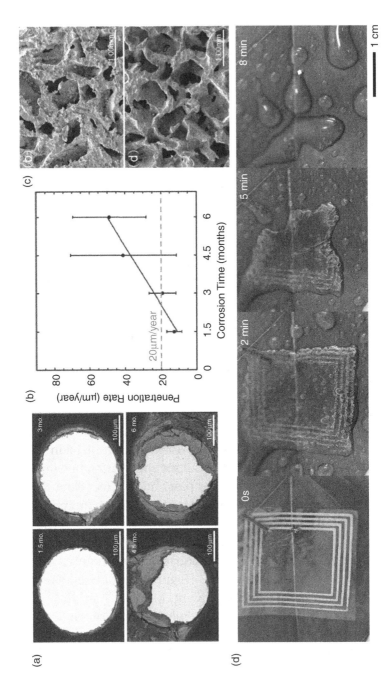

**Figure 4.** Zn as a bioresorbable material. (a) Representative backscattered electron images of cross-sectional areas from Zn implants after being implanted in a Sprague-Dawley rat for 1.5, 3, 4.5, and 6 months

### 2.1.3  Mo

Molybdenum (Mo) is an essential trace element that is needed for at least three enzymes: sulfite oxidase, xanthine oxidase, and aldehyde oxidase. Mo functions as an electron carrier in these enzymes, which catalyze the corresponding reactions with sulfite, purines, pyridines, pyrimidines, and pteridines. Low dietary Mo leads to low urinary and serum uric acid concentrations and excessive xanthine excretion. The Occupational Safety & Health Administration in the USA regulates the maximum Mo exposure in an 8-h working day as 5 mg m$^{-3}$. Chronic exposure to 60–600 mg m$^{-3}$ can cause symptoms including fatigue, headache, and joint pain. Mo is immediately life-threatening at a level of 5000 mg m$^{-3}$.

Mo compounds have low solubility in water, but become quite soluble in water dissolved with oxygen. The dissolution rate of Mo in pH 7 buffer and NaCl solutions is between $10^{-4}$ and $10^{-3}$ μm h$^{-1}$ at room temperature. Such a slow rate is essential for constructing bioresorbable devices that are required to work in a sustained manner in aqueous environments. The dissolution of Mo is influenced by multiple factors, which may lead to effects that contradict

---

**Caption for Figure 4.** (cont.)

(reprinted with permission from Ref. [102], copyright 2013 John Wiley and Sons). (b) Average corrosion rates calculated from measured cross-sectional areas of the Zn implants. The dashed line shows the target value of 20 μm yr$^{-1}$ (reprinted with permission from Ref. [102], copyright 2013 John Wiley and Sons). (c) SEM images of the cross-sectional view of porous Zn scaffolds with different porosities achieved by adjusting the weight ratios between NaCl and Zn during a Zn melt molding process using the NaCl as space holders (reprinted with permission from Ref. [107], copyright 2016 Elsevier Inc.). (d) Dissolution of an NFC coil based on printed Zn microparticles on a leaf under a gentle spray of water (reprinted with permission from Ref. [111], copyright 2014 John Wiley and Sons).

each other. It has been shown that increased oxygen levels from 9 ppm to 37 ppm in DI water can lead to a four-fold increase in the corrosion rate of Mo in such a solution. On the other hand, the presence of ions such as $Na^+$ and $Cl^-$ can reduce oxygen solubility, leading to a decrease in Mo dissolution. Although dissolution rates of Mo thin films are roughly 10 times higher in strong alkaline solution (pH 12) compared to neutral (pH 7) or acidic solutions (pH 2), the induced reductions in oxygen solubility can surpass the effects of pH[38].

Implantable devices made of Mo can function for a longer period of time compared with other transient metals because of its slow dissolution rate. Mo wires (10 μm in thickness) have been used as interfaces to connect a bioresorbable pressure sensor with an external wireless communication system. Mo foil (5 μm in thickness) has been laser cut to form coils and resistive heaters to facilitate implantable drug delivery through a thermally actuated lipid membrane.

## 2.2  Carbon-Based Materials

Allotropes of carbon (C) such as carbon nanotubes (CNTs) and graphene have attracted lots of attention as a result of their unique electrical and mechanical properties. Although lacking systematic studies about the dissolution and biocompatibility of CNTs and graphene, a recent proof-of-concept study has indicated that strong oxidative enzymes may be able to degrade CNTs. Catalytic degradation of carboxylated single-wall carbon nanotubes (SWCNTs) has been demonstrated by the oxidative activity of horseradish peroxidase (HRP) in the presence of low concentrations of $H_2O_2$. It has been also demonstrated that SWCNTs can be biodegraded in phagolysosomal simulating fluid (PSF), which mimics the acidic oxidizing environment typically present in late-stage endosomes and phagolysosomes of macrophages. On the other hand, pristine SWCNTs were resistant when exposed to the same biological oxidative conditions (HRP or PSF), indicating a strong dependence of CNT degradation on surface functionalization[112]. A comparative analysis used carboxylated MWCNTs in HRP and PSF. The structures of the carboxylated MWCNTs were dramatically affected, showing clear partial degradation within 2 months. Although the biological effect of graphene and other carbon nanocompounds is not fully understood, studies have shown that these materials hold great promise in implantable applications. For example, degradation of functionalized SWCNTs *in vitro* did not trigger any toxic effect. In a study using neuron cells, the presence of graphene had no effect on the growth of neurons[113].

Graphene printed on silk membranes has been used to detect target analytes on tooth enamel. The device can selectively detect bacteria through the self-assembly of antimicrobial peptides on graphene[36]. SWNTs and graphene have been fed to silkworms[114]. The silk collected from these silkworms showed a highly developed graphitic structure with enhanced electrical conductivity, increased elongation at break, and increased toughness. This natural feeding strategy paves a new path for . mass production of strengthened silk fibers that can be used as both electrical and structural materials.

### 2.3 Polymers

Bioresorbable polymers can be categorized into synthetic polymers and natural polymers. Degradation of the polymers is typically achieved by hydrolysis of carbonyl functional groups (i.e. esters, amides, and carbonates), resulting in dramatic changes in their chemical structures. In addition, environmental microorganisms that adhere to the surface of the polymer decompose the macromolecules into small molecular debris through the effect of enzyme secretion. Eventually, the debris will be eliminated from the body in forms such as carbon dioxide and water. The relative hydrophilicity or hydrophobicity of polymers governs the uptake of water into the physical structures of the polymers, resulting in more direct interaction between water molecules and the chemical bonds in hydrophilic polymers and thus in faster dissolution compared with hydrophobic polymers.

Bioresorbable polymers typically serve as substrate materials for bioresorbable electronics. Integration of electronic components on

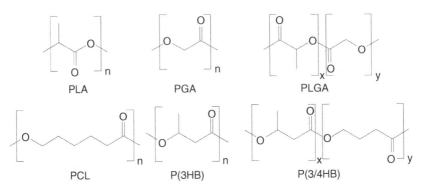

**Figure 5.** Chemical structures of typical aliphatic polyesters
that have been used for making biomedical implants
and bioresorbable electronics.

polymer substrates are realized by transfer printing approaches, in
which components fabricated on donor substrates are transfer
printed onto moisturized polymer substrates with polydimethylsi-
loxane (PDMS) stamps. Separation of the PDMS stamps from the
substrates through drying allows the transferred components to
stay on the substrates.

### 2.3.1  Aliphatic Polyesters

One of the major categories of bioresorbable polymers is aliphatic polyesters, which are synthesized and available in several forms such as polylactic acid (PLA)[115–117], polyglycolic
acid (PGA)[118–120], polylactic-co-glycolic acid (PLGA)[121–123],
poly-ε-caprolactone (PCL)[124–126], and polyhydroxyalkanoate
(PHA)[127–129] (Figure 5). These polymers have been approved
by the US Food and Drug Administration (FDA) for clinical
use[130–132] in tissue engineering and bone repair implants.
Approaches for the fabrication of structures using these polymers include fiber spinning[133, 134], solvent casting[132, 135], melt
molding[136, 137], and gas foaming[138, 139].

Among aliphatic polyesters, PLA is one of the most widely used
biodegradable polymers, with more than 50 yr of use in biomedical

applications; it can be produced either by ring-opening polymerization of lactides or by polymerization of the lactic acid monomers. PLA is classified into two groups, crystalline poly-l-lactic acid (L-PLA or PLLA), which is inert to hydrolysis, and amorphous poly-dl-lactic acid (DL-PLA or PDLA), which is sensitive to hydrolysis. Another common aliphatic polyester is PGA, which is synthesized through the ring opening of the cyclic diesters of glycolic acid. Because of its hydrophilism, PGA has a high degradation rate compared with PLA. Usually, its mechanical strength decreases to 50% of the origin after implantation for 14 days and to 90–95% after 28 days[130]. To achieve a moderate dissolution rate and processability, PLA and PGA are typically copolymerized to yield PLGA, which is most frequently employed for drug delivery[140] owing to its controllable dissolution rate, which is achieved by tuning the crystallinity and the ratio of PLLA, PDLA, and PGA in the copolymers. A more biologically stable polymer is PCL, which has high crystallinity (>45%)[141] and hydrophobic behaviors, offering slower dissolution rates in biological environments compared with PGA and PLGA. Consequently, PCL can be used for long-term implantation. Owing to its good flexibility and machinability, PCL can also be copolymerized with other polymer biomaterials including PLA and PLGA to yield products with unique properties such as shape memory[142-144] and elasticity[145, 146]. Another aliphatic polyester is poly(3-hydroxybutyrate) (P(3HB)), which belongs to a group of polyhydroxyalkanoate (PHA) materials. It is commonly used in surgical sutures[147, 148], soft-tissue repair[149, 150], and drug delivery[151, 152]. However, weaknesses of P(3HB) include brittleness, poor processability, and slow degradation, resulting in the development of poly(3-hydroxybutyrate-co-4-hydroxybutyrate) (P(3/4HB)), whose elasticity can be progressively improved by increasing the content of 4HB to adapt to the requirements of soft tissue regeneration[153].

Major applications of aliphatic polyester include sutures[154], scaffolds[155, 156], fracture fixation[157, 158], drug delivery[159], and stents[160, 161]. They have also been used as substrates for organic thin-film transistors[162, 163], sensors[44, 164], and functional circuits[46], or as surface coatings to extend the implantation period of bioresorbable electronics[45].

### 2.3.2  Polyanhydrides

Polyanhydrides are a class of polymers that contain two carbonyl groups bound together by an ether bond. Unlike aliphatic polyesters, which exhibit bulk erosion patterns, polyanhydrides exhibit surface erosion, leading to predictable hydrolytic degradation ranging from days to months. As a result, polyanhydrides offer much better encapsulation performance compared with other biodegradable polymers, allowing sustained protection of bioresorbable electronic devices embedded within the polyanhydrides. Biocompatibility studies have been conducted both *in vitro* and *in vivo*, generating degradation products that are non-mutagenic and non-cytotoxic.

The degradation of the anhydride bonds in polyanhydrides is highly dependent on the polymer backbone, and can vary by more than six orders of magnitudes. Within different kinds of polyanhydrides, aliphatic homo-polyanhydrides, such as poly-(sebacic anhydride) (poly(SA)) have been found to show rapid degradation. Copolymerization of aliphatic diacid monomers with hydrophobic aromatic diacid monomers or aliphatic fatty acid dimers can slow down the degradation rate. To compensate for the poor mechanical properties of polyanhydrides due to their low molecular weights, methacrylated polyanhydrides have been developed as injectable, crosslinkable biomaterials for tissue engineering[154].

Polyanhydrides have been used as the packaging materials of a complete bioresorbable primary battery[165], which can last 24 h in a liquid electrolyte, showing better stability than other bioresorbable polymer materials. In addition, bioresorbable implantable devices showed a longer working period when packaged with polyanhydrides, owing to the surface erosion properties of polyanhydrides[166].

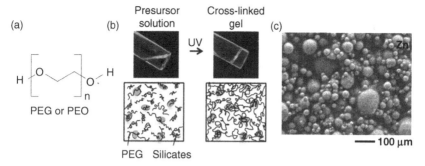

**Figure 6.** Bioresorbable polymer PEG (or PEO). (a) The chemical structure of PEG; (b) PEG-based polymers that show UV crosslinking capability, and can be used to create hydrogel *in situ* from solutions (reprinted with permission from Ref. [177], copyright 2011 Elsevier Inc.); (c) PEO polymers can be used as binders to prepare Zn pastes for printable bioresorbable electronics
(reprinted with permission from Ref. [111],copyright 2014 John Wiley and Sons).

### 2.3.3  Poly(ethylene glycol) (PEG)

Among synthetic polymers, poly(ethylene glycol) (PEG) or poly(ethylene oxide) (PEO) (Figure 6a) are most commonly used for hydrogel fabrication in the biomedical area[167–170], and are currently approved by the FDA for several medical applications[171, 172] including drug delivery, tissue engineering scaffolds, and surface functionalization. Those polymers with molecular weights that are smaller than 100,000 are usually called PEG, while polymers with higher molecular weights are classified as PEO. The rates at which the polymers can be removed from the body are inversely proportional to polymer molecular weight, indicating that PEG can be processed much more easily than PEO in the body. Each end of the PEG chains can be modified with acrylates[173, 174] to allow photo-crosslinking that can yield hydrogels via UV exposure[175–177](Figure 6b). This photoinduced polymer crosslinking method allows hydrogels to be created *in situ*, and offers minimally invasive approaches for tissue replacement or augmentation through endoscopes, catheters, or subcutaneous injections. In addition, thermally reversible hydrogels have been formed from triblock copolymers such as PLLA–PEG–PLLA[178, 179], PCL–PEG–PCL[180, 181] and PLGA–PEG–PLGA[182, 183] to allow the phase transition of the copolymers at room and body temperatures. PEO or PEG are currently used as packing materials for bioresorbable batteries[184] owing to their ability to be thermally cast. Dissolving PEO in organic solvents such as methanol and ethanol at 60–80 °C gives viscous solutions, offering sufficient adhesion to combine bioresorbable substrates together[111]. PEO polymers can also be used as binders to create composite materials such as bioresorbable conductive pastes containing Zn and W microparticles as fillers[111] (Figure 6c).

### 2.3.4   *Protein-Based Natural Polymers*

Proteins are highly complex polymers made out of 21 different amino acids, and are naturally occurring from animals, plants, insects, and fungi. Natural polymers can be derived from proteins such as silk fibroin[185, 186], collagen[187, 188], gelatin[189, 190], elastin[191, 192], and albumin[193]. Major biomedical applications of these polymers include scaffolds[194–196], stents[197], and drug delivery[198, 199]. These proteins can be used either as single components or combined with other bioresorbable polymers (e.g. PLA[200, 201], PLGA[202, 203], and PCL[134, 204]) to yield improved mechanical and degradation properties. Some of these polymers such as silk and gelatin have been used as substrates for bioresorbable electronics, and some (e.g. chicken albumen[193, 205], silk[206, 207], and gelatin[208, 209]) can be used as dielectrics for transistors. In addition, natural proteins may be used to construct dielectric-modulated transistors for sensing specific biomolecules using receptor/ligand systems such as biotin/streptavidin[210, 211], phosphoprotein/zinc(II)-dipicolylamine[212], immunoglobulin E (IgE)/anti-IgE[213], and albumin/anti-albumin[214].

One notable example of a protein-based natural polymer is silk, which is a polypeptide polymer that can be produced in the glands of many arthropods (e.g. silkworms, spiders, scorpions, and mites).

Silk consists of two main proteins: fibroin and sericin, both of which contain the same 18 amino acids, such as glycine, alanine and serine, in different amounts and combinations. Constituently, the chemical structure of silk can be expressed as a recurrent amino acid sequence with glycine, serine, and alanine, i.e. (Gly–Ser–Gly–Ala–Gly–Ala)$_n$ (Figure 7a). Fibroin fibers are encased in sericin[215], and can be separated from sericin by boiling silk in an alkaline solution through a degumming process[216] (Figure 7b). A single cocoon from silkworms yields silk with a length of 600–1500 m, which is much longer than silk from a spider's web (~12 m). Thus, silkworm cocoons are typically used to prepare silk-based biomaterials.

Processed silk after the removal of sericin has a hierarchical structure (Figure 7c), which contains hydrophobic $\beta$-sheets connected with a hydrophilic and semi-amorphous matrix that consists of less orderly $\beta$-structures, $3_1$ helices and $\beta$-turns [217, 218]. Each $\beta$-sheet is made of assembled fibroin fiber sheets formed by laterally connected fibroin fibers through hydrogen bonds. The glycine content allows for tight packing of the $\beta$-sheets, and contributes to its rigid structure and high tensile strength[217].

The dissolution behavior of silk can be controlled by the annealing parameters used to obtain silk sheets. The silk fibroin is water soluble in its $\alpha$-helix and random coil forms. The conformational transition from $\alpha$-helix and random coil to highly stable $\beta$-sheets yields good resistance to dissolution as well as thermal and enzymatic degradation. This transition can be achieved by temperature-controlled water-vapor annealing, in which $\alpha$-helix-dominated silk I structures and highly crystallized $\beta$-sheet-dominated silk II structures can be generated by heating water-based silk solutions at 4 °C and 100 °C, respectively[219]. Silk fibroin is susceptible to enzymatic degradation by proteases such as chymotrypsin[220, 221] and actinase[222], resulting in the production of amino acids, which are easily absorbed.

Owing to its biocompatibility and controlled biodegradability, silk has been employed for manufacturing diverse biomedical implants

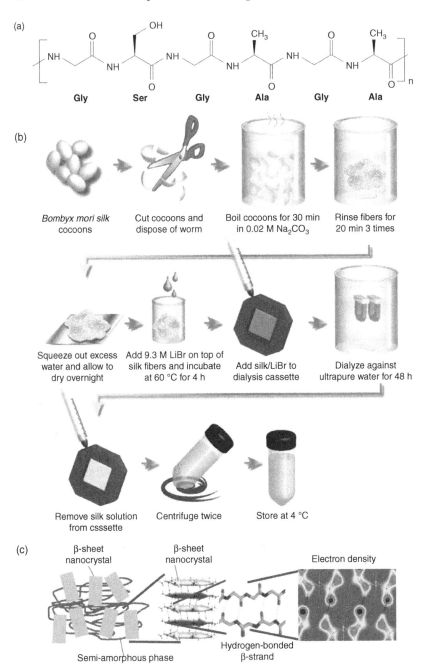

**Figure 7.** Silk as a bioresorbable material. (a) Chemical structure of silk fibroin; (b) The degumming process of silk. This process allows

such as drug carriers[223–225], grafts[226, 227], scaffolds[228, 229], and sutures[230]. In addition, silk has intensive use in bioresorbable electronics as substrates for bioresorbable active and passive components [34, 43, 231]. Fully resorbable circuits with coils, rectification diodes, resistors, and transistors have also been demonstrated using silk substrates[15]. Opto-electric applications of silk involve using silk to form optical gratings[232], wave guides[233], and metamaterials with frequency responses at the terahertz frequency ranges[234].

Another natural polymer from soluble protein compounds is gelatin, which is used commonly for capsules for oral drug ingestion[235, 236]. Gelatin can be obtained by acid or alkaline pre-treatments of collagen, resulting in reduced crosslinks in collagen components[190]. The pre-treated collagen can then be partly hydrolyzed with hot water above 40 °C, followed by treatment with some classic techniques such as settling, filtration, and centrifugation to obtain the final product[237]. The quality of the gelatin depends on pH, temperature, and extraction time used in collagen processing. Parameters such as molecular weight and isoelectric point can be changed depending on the processing conditions. *In vivo* dissolution and degradation of gelatin are more rapid than for silk, because gelatin can readily swell and dissolve in water[238–240]. The process can be facilitated by escalated pH levels[241] and enzymatic degradation[242, 243]. Gelatin has been used as a substrate material for organic field effect transistors[244]. It can also be used as a gate dielectric to offer high device performance

---

**Caption for Figure 7.** (cont.)

the separation of fibroin and sericin, resulting in biodegradable silk solutions that are ready to be applied through spin-coating, molding, and spraying approaches (reprinted with permission from Ref. [216], copyright 2011 Nature Publication Group).
(c) Schematic of the hierarchical silk structure
(reprinted with permission from Ref. [217], copyright 2011 Nature Publication Group).

because of the contribution of negatively charged ions generated by the interaction of water with the polar OH-group in ammonic acids in the gelatin[208].

### 2.3.5  *Bioresorbable Elastomers*

Bioresorbable elastomers have been developed for the demands of curvilinear, stretchable biological tissues and some unique applications that require planar inward forces to facilitate wound or fracture closure[245, 246]. They have been used as grafts[247, 248] and scaffolds[249, 250] for soft tissue regeneration on hearts[251, 252], blood vessels[253, 254], cartilage[255, 256], and nerves[257, 258]. Presently, there are a few resorbable elastomeric polymers including polyurethane (PEU)[259, 260], poly(trimethylene carbonate) (TMC)[261, 262], poly(glycerol sebacate) (PGS)[251, 263], poly-4-hydroxybutyrate (P4HB)[264, 265], and poly(1,8-octanediol citrate) (POC)[250, 266]. Representative bioresorbable elastomers that have been well studied are PGS (Figures 8a and 8b)[267] and POC (Figures 8c and 8d)[268].

The elasticity of PGS is obtained through a covalently cross-linked, three-dimensional network of random coils with hydroxyl groups attached to its backbone. PGS materials offer 267 ± 59.4% stretchability with average Young's modulus in the range of 0.282±0.0250 MPa and ultimate tensile strength larger than 0.5 MPa[269]. These mechanical properties can be tailored by changing the curing temperature, molar ratio of glycerol to sebacic acid, and curing time. In addition, copolymerized PGS with other bioresorbable polymers such as PCL[270] and PEG[267] can yield varied Young's modulus, tensile strength, and stretchability (Figure 8b). The *in vitro* degradation rate of PGS sheets crosslinked at 125 °C in a culture medium is 0.6–0.9 mm month$^{-1}$, while the *in vivo* degradation rate is in the range of 0.2–1.5 mm month$^{-1}$[271]. PGS has been used as a dielectric for fully resorbable pressure sensors based on parallel plate capacitors, and has shown stable performance in a PBS buffer solution for more than 7 weeks[272].

POC is synthesized through a polycondensation reaction between citric acid and linear aliphatic diols (1,8-octanediol).

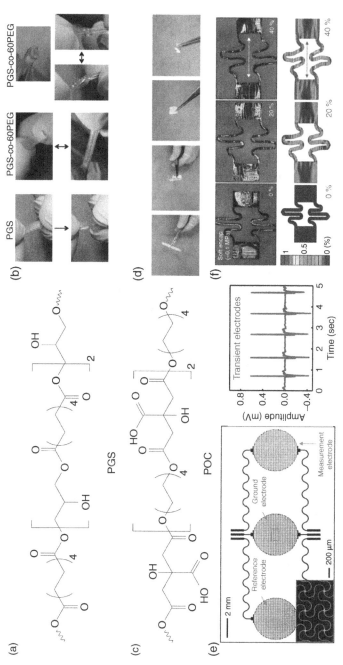

**Figure 8.** Bioresorbable elastomers. (a) The chemical structure of PGS. (b) Comparison of the stretchability of PGS and its copolymers (reprinted with permission from Ref. [267], copyright 2013 Elsevier Inc.). (c) The chemical structure of POC. (d) Demonstration of flexibility of a POC tube (reprinted with permission from Ref. [274], copyright 2006 Elsevier Inc.). (e) A flexible and stretchable skin sensor on a POC substrate. This sensor can be used to conduct biopotential sensing on skin (reprinted with permission from Ref. [275], copyright 2015 American Chemical Society). (f) Microscopic images and FEM results for devices under varied strain levels

(reprinted with permission from Ref. [275], copyright 2015 American Chemical Society).

These elastomers use non-toxic, readily available, and inexpensive citric acid, and incorporate homogeneous biodegradable cross-links to confer elasticity to the resulting material (Figure 8d). The mechanical properties can be tuned by types of diol monomers and processing parameters including post-polymerization time[273] and temperature[274]. For example, increasing post-polymerization time from 1 day to 6 days at 120 °C increases the Young's modulus from 2.84 MPa to 6.44 MPa, while decreasing the stretchability from 253% to 117%[273], and decreasing the post-polymerization temperature from 120 °C to 80 °C leads to an increase in stretchability from ~225% to 375%[274]. Biocompatibility studies using human aortic smooth-muscle cells and endothelial cells show prompt cell confluence and normal phenotype[273]. POC has been used to construct epidermal sensors (Figure 8e) together with Mg electrodes, $SiO_2$ surface coating, and silicon nanomembranes (Si NMs). The resulting devices demonstrate more than 30% stretchability (Figure 8f) with the capability of conducting biopotential and pH sensing on skin[275].

## 2.4   Inorganic Compounds

Inorganic compounds possess small dissolution rates and high chemical stability, and are typically used as surface coating layers to protect bioresorbable implants or as constituent materials for ceramics and composites. The electronics use of inorganic compounds includes gate dielectrics for transistors, spacer layers for capacitors, and insulation layers. Although not listed in the following inorganic bioresorbable materials, it is worth mentioning that zinc oxide, whose biological effects and solubility in biological environments are still under active investigation[276–278], may be used for making light-emitting components[279], gas sensors[280], and mechanical energy harvesters[281] in bioresorbable electronics.

### 2.4.1   Magnesium Oxide (MgO)

MgO can be obtained mostly from calcination of naturally occurring $MgCO_3$ and $Mg(OH)_2$ in minerals and seawater. In neutral and acidic solutions, the dissolution of powdered MgO is associated with the formation of a hydrated intermediate, $Mg(OH)_2$[282]. The overall process of MgO dissolution is controlled by the surface chemical reaction of MgO with $H^+$ ions[283]. The rate of chemical dissolution of MgO accelerates with an increase in both concentrations of $H^+$ ions (from $10^{-4}$ to $10^{-2}$ M) and temperature (from 25 °C to 60 °C). MgO thin films attract significant attention owing to their properties such as high electrical resistivity, high optical transparency, good chemical resistance, excellent thermal and thermodynamic stability, and low refractive index. MgO is an ideal dielectric material for electronics because of its relatively higher permittivity (8–10)[284, 285] as compared with $SiO_2$, and it has been used as gate dielectrics in making transistors[286, 287] and spacer layers of capacitors[288].

### 2.4.2　Silicon Dioxide (SiO₂)

$SiO_2$ is one of the most abundant oxide materials in the Earth's crust. It exists in either amorphous or crystalline forms. $SiO_2$ particles have been used as additives in drugs, cosmetic, and food[289]. $SiO_2$ has also be used in bioresorbable bioglass together with other inorganic compounds such as hydroxylapatite, calcium oxide, and sodium dioxide for bone substitutes[290, 291], cancer treatment[292, 293], and drug delivery[294, 295]. In microelectronics, amorphous $SiO_2$ is used as dielectric (dielectric constant = 3.9) or insulation layers (resistivity = $10^{14}$–$10^{16}$ ohm-cm), and as a passivation material to protect bioresorbable electronic devices from the surrounding aqueous environment.

Dissolution of $SiO_2$ follows the reaction $SiO_2 + 2H_2O \rightarrow H_4SiO_4$[296]. The reaction product of $SiO_2$ hydrolysis is silicic acid $H_4SiO_4$, which is a naturally occurring component in biofluids with a typical concentration in serum ranging from $14 \times 10^{-6}$ to $39 \times 10^{-6}$ $M$[297]. Silicic acid diffuses through the blood stream or lymph, and can be excreted in the urine at a rate of 1.8 mg day$^{-1}$. The reaction to form silicic acid is catalyzed by $OH^-$, hence the hydrolysis rate of $SiO_2$ increases with increasing pH values[298]. Besides the acceleration effects of dissolution by increasing $OH^-$ ions, the kinetics of the dissolution of $SiO_2$ is also influenced by the concentration of other ions in the solutions. As an example, bovine serum (pH ~ 7.4) and seawater (pH ~ 7.8) show rates that are $\approx$9 and $\approx$4 times higher than those observed at similar pH values in buffer solutions, likely as a result of the presence of additional ions (e.g., $K^+$, $Na^+$, $Ca^{2+}$, and $Mg^{2+}$) in these liquids[299]. Furthermore, $SiO_2$ obtained through different processes also exhibits varied dissolution rates. Kang *et al.* studied the dissolution rates of $SiO_2$ obtained through PECVD, electronic beam evaporation, and thermal growth, respectively. The results show that the dissolution rates of the $SiO_2$ in the buffer solution (pH = 7.4 at 37 °C) are on the scale of 10, 0.1, and 0.01 nm day$^{-1}$ corresponding to different film deposition methods, showing significant dependence of the film dissolution on the fabrication methods[299].

### 2.4.3   Silicon Nitride ($Si_3N_4$)

$Si_3N_4$ is a preferred ceramic material known for its chemical stability, high wear resistance, and low friction coefficient. $Si_3N_4$-based ceramics are considered as an excellent alternative material to replace alumina in many applications[300, 301] involving artificial knee joints, hip balls, and acetabula. Several works on biocompatibility and bioactivity of $Si_3N_4$ outline that $Si_3N_4$ -based ceramics can be used as materials for clinical applications in the field of hard tissue surgery. Biological studies of installing $Si_3N_4$ implants into the distal positions of rabbits' tibias reveal osteoconduction processes indicated by the presence of a bone bridge between the implants and the tibias. In addition, newly formed bone exhibits high quality, as shown by nutrient foramens in the bone[302]. An *in vitro* test of cytotoxicity using two $Si_3N_4$-containing ceramic pieces indicates that the presence of $Si_3N_4$ does not cause cell death, showing good biocompatible[303]. Electronic properties of $Si_3N_4$ include high dielectric constant (dielectric constant = 7), large electronic gap, and high resistance against radiation, leading to the extensive use of $Si_3N_4$ in microelectronic devices as gate dielectrics in thin-film transistors[304, 305] and as a charge storage medium in non-volatile memories[306, 307].

Dissolution of $Si_3N_4$ in physiological environments compares favourably with $SiO_2$, which is an intermediate product in the two-step reactions that can eventually turn $Si_3N_4$ into silicic acid. The process of hydrolysis of $Si_3N_4$ can be divided into the following equations: $Si_3N_4 + 6H_2O \rightarrow 3SiO_2 + 4NH_3$ and $SiO_2 + 2H_2O \rightarrow H_4SiO_4$. As a result, the overall reaction can be written as $Si_3N_4 + 12H_2O \rightarrow 3H_4SiO_4 + 4NH_3$. Owing to the generation of $SiO_2$ in the reaction, the dependence of the hydrolysis rate on pH and ion concentrations might be expected to be similar to that observed in $SiO_2$[299].

## 2.5  Semiconductors

Semiconductor materials are significant in achieving active components in bioresorbable electronics. The most frequently used semiconductor material is silicon (Si), which has been used as a scaffold for orthopaedic implants[308], and can be used to support the attachment and growth of a variety of mammalian cells. Another semiconductor material that hydrolyzes in water is germanium (Ge)[309-311], whose biological effects require further study. In addition, bioresorbable organic semiconductors such as β-carotene, indigo, perylenediimide, and indanthrene yellow G may also be used to construct active components in bioresorbable electronics[244].

### 2.5.1  Silicon

Si is a trace element in the human body[312], and is widely available in food[313] and drinking water[314]. Dietary intake of Si is between 20–50 mg day$^{-1}$ for people in the western countries[315], and around 140–204 mg day$^{-1}$ in China and India where plant-based foods are more dominant in the diet[316, 317]. Si is generally considered as a chemically stable material, partly because of the native oxide on its surface. However, at nanoscale thicknesses (i.e., in the form of nanomembranes), even slow reaction and degradation rates can significantly alter the morphology and properties of Si, leading to complete hydrolysis of Si nanomembranes (Si NMs) on the scale of a few days or weeks. The hydrolysis of Si depends critically on the chemical composition of the solution, the temperature, and the doping types and levels of the Si NMs[318]. The hydrolysis of Si is governed by the following chemical reaction: $Si + 4H_2O \rightarrow Si(OH)_4 + 2H_2$. Experimental results from placing $3 \times 3$ cm$^2$ Si NMs (thickness: 70 nm) in phosphate-buffered saline (PBS; pH of 7.4) at both body temperature (37 °C) and room temperature (25 °C) show dissolution rates of 4.5 nm day$^{-1}$ and 2 nm day$^{-1}$, respectively[15]. Yin *et al.* reported a systematic study of the dissolution rates of Si NMs at physiological temperature and pH, and suggested the dependence of dissolution rates on anion concentrations and temperature[319]. The mechanism of Si dissolution has been simulated, and can be possibly connected to chloride and phosphate ions, both of which promote nucleophilic attachment by weakening the bonds of atoms on the Si surface, and prevent formation of surface $SiO_2$. At pH 7.4 and room temperature, dissolution rates of Si in spring water, PBS, and bovine serum were determined to be 0.5 nm day$^{-1}$, 2.5 nm day$^{-1}$, and 35 nm day$^{-1}$, respectively. Raising the temperature to 37 °C promotes dissolution of Si owing to increased mobility of atoms and ions in the solutions. Recent systematic studies of the hydrolysis of mono-Si(100) in solutions with different pH values (between 6 and 10), ion concentrations, and temperatures indicate that the dissolution rates of Si with low or modest doping levels are in the range of 0.5–624 nm day$^{-1}$. These values are well within a range that leads to the complete disappearance of Si NMs (~300 nm or less) on timescales that are relevant for many potential uses of bioresorbable electronics[311].

Toxicity studies of Si NMs have been conducted in live rats, suggesting excellent biocompatibility of Si NMs at the device level[318]. No dissolution residue could be observed by either naked eye or stereomicroscopic analysis after 5-weeks' post-implantation of Si NMs. Hematoxylin and eosin staining and immunohistochemistry of skin sections demonstrated similar levels of immune cells compared with control groups implanted with high-density polyethylene. In addition, rats implanted with

Si NMs showed no significant body weight loss compared with rats without any implant.

### 2.5.2 Bioresorbable Organic Semiconductors

Organic semiconductors have also been used to make active devices. Many bioresorbable organic semiconductor materials are derived from natural pigments such as $\beta$-carotene[320] and chlorophyll[321, 322] or from dyes such as indanthrene yellow G[323], brilliant orange RF[323], perylenediimide[324, 325], and indigo[326], which are typically used as food, textile, and cosmetic colorants[327]. These organic semiconductors contain π-conjugated carbon back-bones that allow delocalization of electrons. Some of these materials have nutritional benefits to the body; for example, $\beta$-carotene can be converted to vitamin A by dioxygenase, and chlorophyll can supply Mg that maintains muscle and bone health and chlorophyllin that is used for the treatment of certain types of cancers. Some organic materials such as $\beta$-carotene, chlorophyll, and indigo typically have low water solubility[328, 329], and can be degraded within the body to yield dissolvable products[330]. The utilization of these semiconductors in making organic thin-film transistors has been explored by many researchers[244]; however, these natural organic semiconductors show intrinsic low mobility (~$4 \times 10^{-4}$ cm$^2$ V$^{-1}$ s$^{-1}$ for $\beta$-carotene and ~$2 \times 10^{-4}$ cm$^2$ V$^{-1}$ s$^{-1}$ for indigo)[323] that is much less than many non-bioresorbable semiconductor materials such as C$_{60}$ and pentacene[331]. Significant efforts are required to improve the electrical properties of bioresorbable organic semiconductors to yield high-performance electronics for realistic applications.

## 3 Fabrication of Bioresorbable Electronics

Fabrication of transient devices involves complex processes that contain major challenges related to the properties of bioresorbable materials. First, bioresorbable materials are subject to the influence of hydroxyl, oxygen, and nitrogen, the presence of which in the surrounding environment will trigger chemical reactions and degradation of the materials. Second, some polymeric bioresorbable materials suffer significant phase and morphology changes at elevated temperatures (above 100 °C), and, thus, are not suitable for fabrication methods that involve excessive heat generation. As a result, transient electronic devices are predominately achieved by complex, time-consuming processes based on anhydrous surface micromachining and CMOS fabrication methods on regular substrates followed by transfer printing to bioresorbable substrates.

A typical fabrication process of transient electronics starts with the spin-coating of poly(methyl methacrylate) (PMMA) as a sacrificial layer and a diluted polyimide (Di-PI) layer on a silicon substrate, followed by transfer printing and patterning of a Si nanomembrane with doped regions predefined on a silicon-on-insulator (SOI) wafer. A photoresist is then used to define patterns for transient metal deposition. A transient metal layer can then be deposited through electron-beam (ebeam) evaporation. The photoresist layer and the metal layer on the photoresist are then lifted off in acetone, leaving traces of transient metal that connects both active and passive components. Dielectric layers made of inorganic compound such as MgO and $SiO_2$ can then be deposited through physical or chemical deposition, followed by reactive plasma etching using a patterned photoresist layer as a mask. Repeated processes of metal deposition, lifting off, dielectric deposition, and etching lead to stacked transient patterns. Eventually, the transient devices can be released in acetone by removal of PMMA and picked up by a PDMS stamp. The Di-PI layer is eventually removed through reactive ion. etching (RIE). Transfer printing of the resulting transient electronic device to a transient substrate made of a bioresorbable polymer completes the fabrication process. In addition, when the requirement of linewidth is not critical, transient patterns can be formed by deposition through a stencil shadow mask, which blocks the deposition of evaporated transient materials on defined regions. The above-mentioned fabrication process can all be conducted by using chemicals that do not contain water, adding cost and time of the fabrication with decreased yield. As a result, although the benefits of using bioresorbable electronic devices have been well acknowledged, they have not yet been commercialized to generate a broader social impact through innovative health care products and environmentally friendly circuits. The primary reason is the largely unexplored mass manufacturing technology that can yield high-quality and low-cost bioresorbable electronic devices.

Recent efforts in developing new fabrication techniques of bioresorbable electronics have led to devices based on direct laser cutting[46], aerosol or inkjet printing[332], screen printing[111], and evaporation vapor condensation[333], demonstrating cost-effective and rapid-prototyping approaches compared with the traditional CMOS fabrication process. Many of these techniques demand the use of micro/nano particles that are mixed with organic solvent, binders, and surfactants. Preprocessing of the transient particle using ball milling may yield irregular shapes of transient nanoparticles that offer more contact points and an enhanced tunneling effect, leading to increased possibilities of achieving higher conductivity. In addition, post-processing using localized heat treatment through photonic sintering can induce the formation of a conductive matrix of these printed nanoparticles, resulting in improved mechanical strength and conductivity of the sintered patterns. It is believed that with the addition of carbon-based materials that offer either metallic or semiconductive properties, fully transient devices may be achieved by printable technology.

## 4  Applications of Bioresorbable Materials in Electronics

Bioresorbable materials have been used to make electronic devices that work either as single components or in simple systems. It has been reported that electronic components can be combined to yield complex systems that may eventually lead to fully implantable systems based completely on bioresorbable materials.

Bioresorbable batteries[165, 184, 334, 335] are the fundamental components in bioresorbable electronics for supplying power to the implanted bioresorbable systems. Yin *et al.* presented bioresorbable primary batteries that contain stacked Mg–Mo cells (Mg foils as the anodes and Mo as the cathodes) in a polyanhydride package filled with PBS as electrolyte. A single Mg–Mo cell with 1 cm$^2$ active area offers a capacity of $\approx$2.4 mAh (0.1 mA cm$^{-2}$ for 24 h) at a stable voltage output of ($\approx$0.4–0.7 V), corresponding to a specific capacity $\approx$276 mAh g$^{-1}$ (normalized with anode mass). Four stacked Mg–Mo cells with a dimension of 3 cm $\times$ 1.3 cm $\times$ 1.6 cm (Figure 9a) and a weight of 3.5 g yield a constant current density (0.1 mA cm$^{-2}$) at a stable voltage output $\approx$1.5–1.6 V for up to 6 h[165] (Figure 9b). A more miniaturized primary battery has been developed based on PCL-coated Mg/Fe cells. The battery occupies a volume of less than 0.02 cm$^3$, and offers an average power of approximately 30 μW for 100 h, sufficient for powering a commercial pacemaker for 4 days[336]. An alternative polymer electrolyte is used by Jia *et al.* in a ultra-compact magnesium–air battery[335]. The polymer electrolyte is made of choline nitrate, which is a gel-like compound that offers mechanical robustness and high ionic conductivity (8.9 $\times$ 10$^{-3}$ S cm$^{-1}$). The electrolyte is embedded in a chitosan package sandwiched by polypyrrole–para(toluene sulfonic acid) as the cathode and a Mg alloy anode, resulting in a battery with overall thickness of only 300 μm and a maximum volumetric power density of 3.9 W L$^{-1}$. A rechargeable transient power source has been demonstrated by Douglas *et al.* [184]. This battery uses a gel-based PVA/0.5 M LiClO$_4$ as an electrolyte filled within a PEO spacer, which is sandwiched by vanadium oxide (VOx)-coated porous Si plates (Figure 9c). This battery has a stable specific capacitance of ~21 F g$^{-1}$ according to the charge–discharge curves in Figure 9d. The disintegration and dissolution of the battery can be externally triggered by alkaline solutions (Figure 9e), in which the thin coating of the VOx near the base of the Si plates dissolves away more rapidly than the thicker coating near the top of the materials, resulting in rapid disintegration of VOx from the porous Si and deactivation of the energy-storage function in a few seconds followed by complete device dissolution in 30 min.

Bioresorbable sensors are essential components for bioresorbable electronics, and have been realized in formats such as biopotential electrodes[164, 166], pressure sensors[272, 337, 338], motion sensors[166], and temperature sensors[166]. Because of low

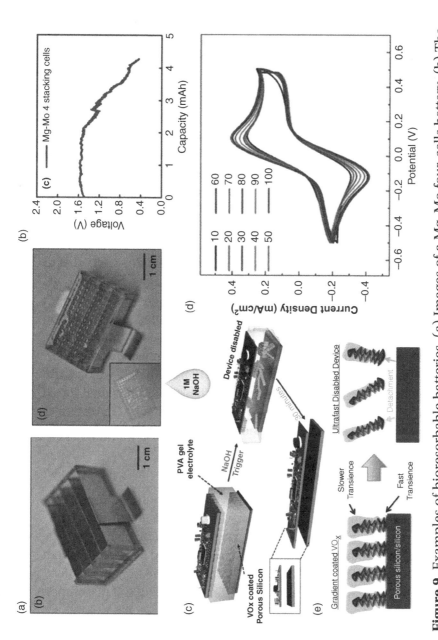

**Figure 9.** Examples of bioresorbable batteries. (a) Images of a Mg–Mo four-cells battery. (b) The discharging behavior of the Mg--Mo cell battery (reprinted with permission from Ref. [165], copyright 2014

dissolution rates, Si is typically used as the electrode materials for implantable transient sensors. Yu *et al.* developed neural interfaces using Si as electrodes equipped with multiplexing capabilities to allow biopotential sensing from the cortical surface and subgaleal space on the cerebral cortex of mice for acute (~h) and chronic (up to 33 days) use[164]. Representative devices include a low-density four-probe sensor (Figure 10a) that was used for measuring the biopotential on the cortical surface of mice (Figure 10b), and a high-density multiplexing sensor containing 64 electrodes and 128 MOSFETs for microscale electrocorticography on the subgaleal space. The active components and electrodes are realized by Si NMs that are interconnected by Mo and passivated by stacked layers of $SiO_2$ and $Si_3N_4$ on PLGA substrates. Kang *et al.* also developed multifunctional Si sensors for the brain[166] (Figure 10c). The sensors contain Si NMs patterned into meander structures to monitor temperature, acceleration, pressure, flow rate, and pH values, which are measured as resistance or conductance changes of the Si NMs (Figure 10d). The Si NMs are sealed within $SiO_2$ on PLGA substrates, and can perform sensing functions up to 30 min. To tackle the issues of power supply and wire connection in implantable devices, Luo *et al.* presented a bioresorbable wireless RF MEMS pressure sensor that remained stable and functional for 86 h after immersing in saline[338]. The sensor used a Zn/Fe bilayer as the conducting material and poly-L-lactide (PLLA) and PCL as

---

**Caption for Figure 9.** (cont.)

John Wiley and Sons). (c) Schematic of the integration of a transient energy-storage device with Si-based electronics. (d) Voltammetric cycling performance for the transient energy-storage device with 19 nm VOx coating on porous Si. (e) Triggered system dissolution by 1 M NaOH aqueous solution. The triggering effect leads to immediate disablement of the device, and full dissolution within 30 min
(reprinted with permission from Ref. [184], copyright 2016 Royal Society of Chemistry).

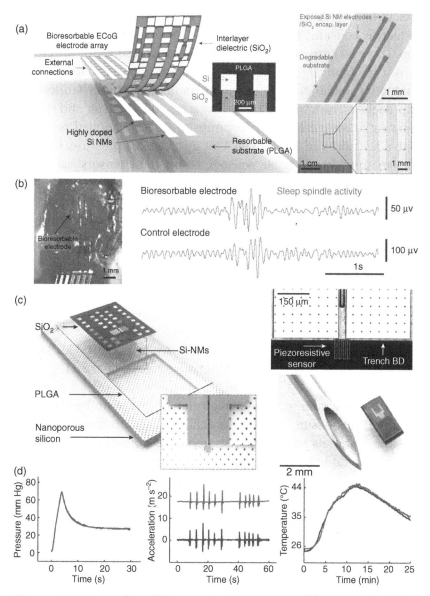

**Figure 10.** Examples of bioresorbable sensors. (a) A thin, flexible neural electrode array with fully bioresorbable construction based on Si NMs as the conducting component. (b) *In vivo* neural recordings in a rat using a passive, bioresorbable electrode array

dielectric and structural materials. The Zn/Fe bilayer forms an inductive coil and parallel capacitive plates sandwiching an air gap. To avoid corrosion from the chemicals and solvents, the entire sensor is fabricated using embossing, multilayer folding, and lamination techniques rather than traditional MEMS fabrication approaches. The fabricated sensor was tested in both air and saline, showing a linear frequency response to external applied pressure with a sensitivity of 39 kHz kPa$^{-1}$ for up to 20 kPa pressure. Besides implantable bioresorbable sensors, edible food sensors have also been presented. Tao *et al.* presented edible sensors that contain antennas and resonators, whose frequency response can be varied with food qualities[339]. These sensors can be placed on the surface of fruits, vegetables, and eggs, to measure their freshness and ripeness.

Both inorganic and organic active components have been developed and used to realize functions such as amplification, multiplexing, radio-frequency generation, and power harvesting. The majority of inorganic active components are based on Si NMs, which are typically processed on SOI wafers (Figure 11a) and transfer printed onto bioresorbable substrates (Figure 11b) such as silk[15, 340], PLGA[164], and thin metal foils[341], allowing either discrete components or high-density arrays to be achieved (Figure 11c). Besides using inorganic semiconductor materials, organic materials have demonstrated their potential as dielectric and semiconductor materials. Guo *et al.* presented a low-voltage

---

**Caption for Figure 10.** (cont.)

(reprinted with permission from Ref. [164], copyright 2016 Nature Publication Group). (c) Schematic illustration of a bioresorbable pressure sensor and its size comparison with an injection needle. (d) Comparison of experiment data of pressure, acceleration, and temperature sensing between a bioresorbable pressure sensor (red) and commercial sensors (blue)

(reprinted with permission from Ref. [166], copyright 2016 Nature Publication Group).

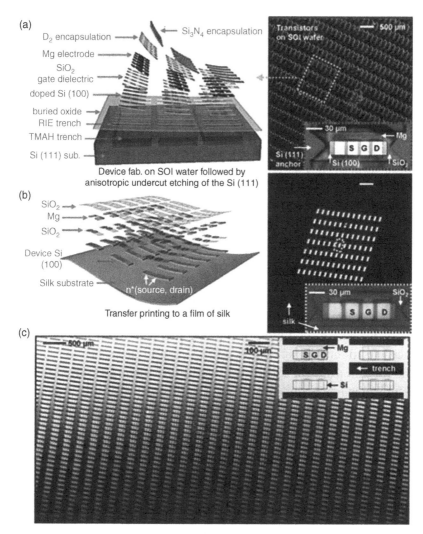

**Figure 11.** Fabrication of transient active components on specialized SOI substrates and their subsequent transfer printing to silk. (a) Schematic exploded view illustration and optical microscope images of devices after completing the fabrication process on a SOI wafer. (b) Schematic illustration and an image of the same devices after transfer printing to a silk film. (c) An image of MOSFET arrays fabricated using a SOI substrates and the transfer printing method (reprinted with permission from Ref. [340], copyright 2013 John Wiley and Sons).

junctionless transient transistor with a sodium alginate membrane both as a substrate and a dielectric layer, and a single patterned Al: ZnO (AZO) thin film to form source/drain electrodes and a channel region. The transistors can be operated at a low voltage of 1 V, and show completely physical transient behavior within 60 min in deionized water[342]. Capelli *et al.* demonstrated OFETs and organic light emitting transistors (OLETs) using both n-type (perylene) and p-type (thiophene) organic semiconductors and silk fibroin that served as dielectrics. The charge mobility and on/off ratio of the bioresorbable OFETs are measured to be on the order $10^{-2}$ cm$^2$ V$^{-1}$ s$^{-1}$ and $10^4$, respectively, while the n-type OLETs offer light emission of ~100 nW[343]. Other promising natural organic semiconductor materials such as vat yellow 1, vat orange 3, $\beta$-carotene, and indigo can be used to make all nature organic electronics, including inverters, ring oscillators, and memory elements[323, 344]. For example, Irimia-Vlad *et al.* have present OFETs and inverters based on indigo[326]. The OFETs were fabricated on substrates made of natural shellac, which can be produced from female lac beetles and trees, and drop-casted to form smooth and uniform substrates. Aluminum oxide (45 nm in thickness) is used as gates covered by a thin layer of tetratetracontane (30 nm in thickness). The threshold voltages of indigo-based OFETs measured with positive source/drain voltages are in the range of –1.5 to –3 V for holes and 4.5 to 7 V for electrons. The electron and hole field-effect mobilities are around $1 \times 10^{-2}$ cm$^2$ V$^{-1}$ s$^{-1}$ for electrons and $5 \times 10^{-3}$ to $1 \times 10^{-2}$ cm$^2$ V$^{-1}$ s$^{-1}$ for holes.

The components mentioned above can be used to construct simple functional bioresorbable systems. A typically feature of these systems is that they handle the issue of power supply through introducing inductive coupling for power transfer through the skin. Hwang *et al.* have developed a pioneering system that contains various components ranging from inductors, capacitors, resistors, diodes, transistors, conductive interconnects, and dielectrics on silk substrates[15]. This conceptual system can be used to harvest environmental electromagnetic fields and convert the RF energy to DC power. This work present a wide range of bioresorbable materials including Mg, Si NMs, MgO, SiO$_2$, and silk, as well as initial studies of the dissolution behavior of these materials, and has inspired many succeeding systems with realistic biomedical applications. Song *et al.* have developed a bioresorbable electronic stent integrated with comprehensive functions such as flow sensing, temperature monitoring, data storage, wireless power/data transmission, inflammation suppression, localized drug delivery, and hyperthermia therapy (Figure 12a). This multifunctional stent contains a Mg-alloy-based stent integrated with metallic temperature and flow sensors and a resistive random access memory array for data storage. The entire device is coated with ceria nanoparticles for anti-inflammation and gold nanorod core/ mesoporous silica nanoparticles for drug delivery through the photothermic effect. When needed, the stent can be used as an antenna for wireless power harvesting and communication[45]. Lee *et al.* have reported a device that combined temperature-sensitive lipid films with electronically programmable, frequency-multiplexed wireless hardware. The device contains four sets of conductive coils connected with meander resistors on PLGA substrates covered with drug molecules as well as lipid membranes (Figure 12b). The coils can respond to an electromagnetic field in different frequencies and drive corresponding resistors to generate heat that can stimulate drug releasing due to phase changes of the lipid membranes, allowing the drug molecules underneath to diffuse through[46](Figure 12c). Besides controllable drug releasing, wireless-powered resistive microheaters can also be directly used as a tool for thermal therapy to control surgical site infection[70].

## 5 Conclusions and Perspective

Materials and devices presented above offer a variety of options to construct fully bioresorbable electronic systems, and have inspired the development of many bioresorbable devices in the form of single components and simple functional systems. Despite great

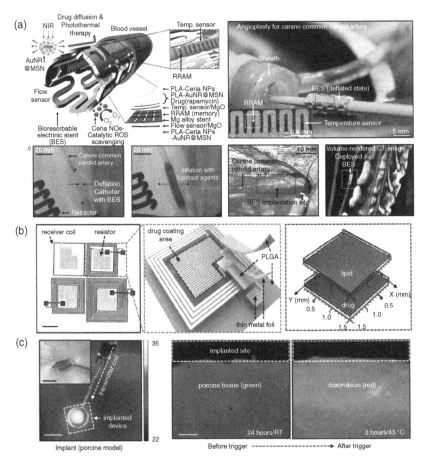

**Figure 12.** Examples of simple functional bioresorbable electronic systems. (a) A bioresorbable electronic stent that contains bioresorbable temperature/flow sensors, memory modules, and bioresorbable/bioinert therapeutic nanoparticles. This stent can be deployed into artery through a balloon catheter (reprinted with permission from Ref. [45], copyright 2015 American Chemical Society). (b) An optical image and schematic exploded view of a device that consists of a 2 × 2 array of inductive coupling Mg coils and serpentine thermal actuators on a PLGA substrates. The top surface of the device is covered by a lipid membrane loaded with drug molecules. (c) *In vivo* operation of a device in a porcine

achievements in bioresorbable electronics, some critical challenges in the development of bioresorbable electronics still require further research efforts to achieve advanced solutions.

One issue is the complex and time-consuming anhydrous fabrication approaches of bioresorbable electronics. Current fabrication processes of bioresorbable electronics typically involve layer-by-layer depositing and patterning of bioresorbable materials on donator substrates that are water insoluble, releasing in solvent, and transfer printing to bioresorbable substrates. These complex fabrication processes result in low yield and high cost when fabricating comprehensive and complicated bioresorbable electronic systems. Exploration of innovative fabrication methods designed specifically for bioresorbable electronics is still ongoing, and may include several potential approaches such as direct printing[345-347], laser processes[348-350], and photonic sintering[351-353].

In addition, the mechanisms for realizing the physical transience of these bioresorbable electronics still largely rely on spontaneous dissolution of surface coating and composition materials in aqueous environments. As the dissolution of bioresorbable materials depends critically on process parameters and physiological conditions, all of which are difficult to be precisely controlled, prediction of device life span and dissolution behaviors becomes extremely difficult. However, triggered transience allows packing of the devices in well-defined and enhanced packages with slow and stable dissolution, while offering rapid deactivation of device

---

**Caption for Figure 12.** (cont.)

model. The device is operated in a triggered mode, which uses heat to induce phase change of a lipid layer. A thermal image indicates the device's heating in the heater region, and the fluorescent optical images on the right indicate drug releasing before and after triggering

(reprinted with permission from Ref. [46], copyright 2015 Nature Publication Group).

functions and breakage of external packages through heat, light, force, or electrical stimulation, leading to more controllable device dissolution. Thus triggered transience can be introduced in future developments of bioresorbable electronics, allowing the bioresorbable devices to work in a stable manner for defined periods of time and dissolve or disintegrate instantly to deactivate the device functions.

Furthermore, the performance of the bioresorbable devices depends critically on the surface packaging or encapsulation materials, which isolate the bioresorbable devices from the external aquatic environment, while possessing tunable dissolution rates. One strategy may be to use surface erosion polymers such as polyanhydrides, which dissolve in water layer by layer. Another approach may be to use stacked inorganic compounds such as $SiO_2$, $Si_3N_4$, and MgO, which are deposited on the bioresorbable electronic devices using pin-hole free-deposition methods such as plasma enhanced chemical vapor deposition and atomic layer deposition. Moreover, packages based on transient metal films/foils offer promising approaches to prevent water molecules from directly contacting with bioresorbable devices.

It is also noticeable that current materials used in bioresorbable electronic devices are mostly limited to some common materials such as Mg, silk, Si, MgO, and $SiO_2$. Many bioresorbable materials such as alloys, composites, and ceramics have demonstrated their use in biomedical implants, and their capabilities in improving mechanical and chemical properties of the implants. However, they are current largely unexplored in the context of bioresorbable electronics. It is expected that methods for integrating various bioresorbable materials may emerge with the increasing effort in developing revolutionary fabrication processes.

The development of bioresorbable electronic technology benefits from an understanding of bioresorbable materials and the knowledge accumulated for development of bioresorbable implants, and offers innovative solutions to eliminate the need for secondary surgeries and to minimize the generation of electronic waste. Advances in processing methods of bioresorbable materials and triggered transience approaches, as well as other relevant techniques, may eventually lead to the appearance of more fully implantable, self-supported bioresorbable electronics systems that are clinically and commercially available to improve societal health levels and reduce environmental pollution.

# References

[1] R. C. Eberhart, S. H. Su, K. T. Nguyen, M. Zilberman, L. Tang, K. D. Nelson, and P. Frenkel, "Bioresorbable polymeric stents: current status and future promise," *J Biomater Sci Polym Ed*, **14**(4), pp. 299–312, 2003.

[2] J. A. Ormiston, P. W. Serruys, E. Regar, D. Dudek, L. Thuesen, M. W. I. Webster, Y. Onuma, H. M. Garcia-Garcia, R. McGreevy, and S. Veldhof, "A bioabsorbable everolimus-eluting coronary stent system for patients with single de-novo coronary artery lesions (ABSORB): a prospective open-label trial," *The Lancet*, **371**(9616), pp. 899–907, 2008.

[3] J. Simon, J. Ricci, and P. Di Cesare, "Bioresorbable fracture fixation in orthopedics: a comprehensive review. Part I. Basic science and pre-clinical studies," *American Journal of Orthopedics (Belle Mead, NJ)*, **26**(10), pp. 665–671, 1997.

[4] L. Ylikontiola, K. Sundqvuist, G. K. Sandor, P. Törmälä, and N. Ashammakhi, "Self-reinforced bioresorbable poly-L/DL-lactide [SR-P (L/DL) LA] 70/30 miniplates and miniscrews are reliable for fixation of anterior mandibular fractures: a pilot study," *Oral Surgery, Oral Medicine, Oral Pathology, Oral Radiology, and Endodontology*, **97**(3), pp. 312–317, 2004.

[5] B. D. Gogas, V. Farooq, Y. Onuma, and P. W. Serruys, "The ABSORB bioresorbable vascular scaffold: an evolution or revolution in interventional cardiology?," *Hellenic J Cardiol*, **53**(4), pp. 301–309, 2012.

[6] S. A. Abbah, C. X. L. Lam, D. W. Hutmacher, J. C. H. Goh, and H.-K. Wong, "Biological performance of a polycaprolactone-based scaffold used as fusion cage device in a large animal model of spinal reconstructive surgery," *Biomaterials*, **30**(28), pp. 5086–5093, 2009.

[7] J. Gresser, K.-U. Lewandrowski, D. Trantolo, D. Wise, and Y.-Y. Hsu, "Soluble Calcium Salts in Bioresorbable Bone Grafts," in *Biomaterials Engineering and Devices: Human Applications*, D. Wise, D. Trantolo,

K.-U. Lewandrowski, J. Gresser, M. Cattaneo, and M. Yaszemski, Eds., Humana Press, 2000, pp. 171–188.

[8] C. Bergmann, M. Lindner, W. Zhang, K. Koczur, A. Kirsten, R. Telle, and H. Fischer, "3D printing of bone substitute implants using calcium phosphate and bioactive glasses," *Journal of the European Ceramic Society*, **30**(12), pp. 2563–2567, 2010.

[9] G. D. Guerra, P. Cerrai, M. Tricoli, S. Maltinti, I. Anguillesi, and N. Barbani, "Fibers by bioresorbable poly(ester-ether-ester)s as potential suture threads: a preliminary investigation," *J Mater Sci Mater Med*, **10**(10/11), pp. 659–662, 1999.

[10] D. Schranz, P. Zartner, I. Michel-Behnke, and H. Akintürk, "Bioabsorbable metal stents for percutaneous treatment of critical recoarctation of the aorta in a newborn," *Catheterization and Cardiovascular Interventions*, **67**(5), pp. 671–673, 2006.

[11] M. A. Woodruff and D. W. Hutmacher, "The return of a forgotten polymer – Polycaprolactone in the 21st century," *Progress in Polymer Science*, **35**(10), pp. 1217–1256, 2010.

[12] R. Auras, B. Harte, and S. Selke, "An overview of polylactides as packaging materials," *Macromolecular Bioscience*, **4**(9), pp. 835–864, 2004.

[13] X. Pang, X. Zhuang, Z. Tang, and X. Chen, "Polylactic acid (PLA): Research, development and industrialization," *Biotechnology Journal*, **5**(11), pp. 1125–1136, 2010.

[14] V. G. Kadajji and G. V. Betageri, "Water soluble polymers for pharmaceutical applications," *Polymers*, **3**(4), pp. 1972–2009, 2011.

[15] S.-W. Hwang, H. Tao, D.-H. Kim, H. Cheng, J.-K. Song, E. Rill, M. A. Brenckle, B. Panilaitis, S. M. Won, Y.-S. Kim, Y. M. Song, K. J. Yu, A. Ameen, R. Li, Y. Su, M. Yang, D. L. Kaplan, M. R. Zakin, M. J. Slepian, Y. Huang, F. G. Omenetto, and J. A. Rogers, "A physically transient form of silicon electronics," *Science*, **337**(6102), pp. 1640–1644, 2012.

[16] Y. Okazaki and E. Gotoh, "Metal release from stainless steel, Co–Cr–Mo–Ni–Fe and Ni–Ti alloys in vascular implants," *Corrosion Science*, **50**(12), pp. 3429–3438, 2008.

[17] A. H. Greene, J. D. Bumgardner, Y. Yang, J. Moseley, and W. O. Haggard, "Chitosan-coated stainless steel screws for fixation in contaminated fractures," *Clinical Orthopaedics and Related Research*, **466**(7), pp. 1699–1704, 2008.

[18] F. Nie, S. Wang, Y. Wang, S. Wei, and Y. Zheng, "Comparative study on corrosion resistance and in vitro biocompatibility of bulk nano-crystalline and microcrystalline biomedical 304 stainless steel," *Dental Materials*, **27**(7), pp. 677–683, 2011.

[19] M. Grądzka-Dahlke, J. Dąbrowski, and B. Dąbrowski, "Modification of mechanical properties of sintered implant materials on the base of Co–Cr–Mo alloy," *Journal of Materials Processing Technology*, **204**(1), pp. 199–205, 2008.

[20] K. Teigen and A. Jokstad, "Dental implant suprastructures using cobalt–chromium alloy compared with gold alloy framework veneered with ceramic or acrylic resin: a retrospective cohort study up to 18 years," *Clinical Oral Implants Research*, **23**(7), pp. 853–860, 2012.

[21] W.-C. Witzleb, J. Ziegler, F. Krummenauer, V. Neumeister, and K.-P. Guenther, "Exposure to chromium, cobalt and molybdenum from metal-on-metal total hip replacement and hip resurfacing arthro-plasty," *Acta Orthopaedica*, **77**(5), pp. 697–705, 2006.

[22] X. Liu, P. K. Chu, and C. Ding, "Surface modification of titanium, titanium alloys, and related materials for biomedical applications," *Materials Science and Engineering: R: Reports*, **47**(3), pp. 49–121, 2004.

[23] C. Elias, J. Lima, R. Valiev, and M. Meyers, "Biomedical applications of titanium and its alloys," *JOM*, **60**(3), pp. 46–49, 2008.

[24] M. Geetha, A. Singh, R. Asokamani, and A. Gogia, "Ti based bioma-terials, the ultimate choice for orthopaedic implants-a review," *Progress in Materials Science*, **54**(3), pp. 397–425, 2009.

[25] A. Godara, D. Raabe, and S. Green, "The influence of sterilization processes on the micromechanical properties of carbon fiber-rein-forced PEEK composites for bone implant applications," *Acta Biomaterialia*, **3**(2), pp. 209–220, 2007.

[26] J. W. Thomas, C. W. Michael, L. M. Janice, L. P. Rachel, and U. E. Jeremiah, "Nano-biotechnology: carbon nanofibres as improved neural and orthopaedic implants," *Nanotechnology*, **15**(1), p. 48, 2004.

[27] D. Adams, D. F. Williams, and J. Hill, "Carbon fiber-reinforced carbon as a potential implant material," *Journal of Biomedical Materials Research*, **12**(1), pp. 35–42, 1978.

[28] A. Dalal, V. Pawar, K. McAllister, C. Weaver, and N. J. Hallab, "Orthopedic implant cobalt-alloy particles produce greater toxicity

and inflammatory cytokines than titanium alloy and zirconium alloy-based particles in vitro, in human osteoblasts, fibroblasts, and macrophages," *Journal of Biomedical Materials Research Part A*, **100**(8), pp. 2147–2158, 2012.

[29] M. G. Shettlemore and K. J. Bundy, "Toxicity measurement of orthopedic implant alloy degradation products using a bioluminescent bacterial assay," *J Biomed Mater Res*, **45**(4), pp. 395–403, 1999.

[30] C. M. George, D. R. Howard, I. L. Allan, G. Claire-Anne, E. G. Robert, and V. B. Ravi, "Implanted neural electrodes cause chronic, local inflammation that is correlated with local neurodegeneration," *Journal of Neural Engineering*, **6**(5), p. 056003, 2009.

[31] N. J. Hallab, B. W. Cunningham, and J. J. Jacobs, "Spinal implant debris-induced osteolysis," *Spine*, **28**(20S), pp. S125–S138, 2003.

[32] K. Muller and E. Valentine-Thon, "Hypersensitivity to titanium: clinical and laboratory evidence," *Neuro Endocrinol Lett*, **27** Suppl. 1, pp. 31–35, 2006.

[33] N. J. Hallab and J. J. Jacobs, "Biologic effects of implant debris," *Bulletin of the NYU Hospital for Joint Diseases*, **67**(2), p. 182, 2009.

[34] D.-H. Kim, Y.-S. Kim, J. Amsden, B. Panilaitis, D. L. Kaplan, F. G. Omenetto, M. R. Zakin, and J. A. Rogers, "Silicon electronics on silk as a path to bioresorbable, implantable devices," *Applied Physics Letters*, **95**(13), p. 133701, 2009.

[35] D.-H. Kim, J. Viventi, J. J. Amsden, J. Xiao, L. Vigeland, Y.-S. Kim, J. A. Blanco, B. Panilaitis, E. S. Frechette, D. Contreras, D. L. Kaplan, F. G. Omenetto, Y. Huang, K.-C. Hwang, M. R. Zakin, B. Litt, and J. A. Rogers, "Dissolvable films of silk fibroin for ultrathin conformal bio-integrated electronics," *Nat Mater*, **9**(6), pp. 511–517, 2010.

[36] M. S. Mannoor, H. Tao, J. D. Clayton, A. Sengupta, D. L. Kaplan, R. R. Naik, N. Verma, F. G. Omenetto, and M. C. McAlpine, "Graphene-based wireless bacteria detection on tooth enamel," *Nat Commun*, **3**, p. 763, 2012.

[37] B. D. Lawrence, M. Cronin-Golomb, I. Georgakoudi, D. L. Kaplan, and F. G. Omenetto, "Bioactive silk protein biomaterial systems for optical devices," *Biomacromolecules*, **9**(4), pp. 1214–1220, 2008.

[38] L. Yin, H. Cheng, S. Mao, R. Haasch, Y. Liu, X. Xie, S.-W. Hwang, H. Jain, S.-K. Kang, Y. Su, R. Li, Y. Huang, and J. A. Rogers, "Dissolvable metals for transient electronics," *Advanced Functional Materials*, **24**(5), pp. 645–658, 2014.

[39] H. K. Makadia and S. J. Siegel, "Poly lactic-co-glycolic acid (PLGA) as biodegradable controlled drug delivery carrier," *Polymers (Basel)*, **3**(3), pp. 1377–1397, 2011.

[40] S.-K. Kang, S.-W. Hwang, H. Cheng, S. Yu, B. H. Kim, J.-H. Kim, Y. Huang, and J. A. Rogers, "Dissolution behaviors and applications of silicon oxides and nitrides in transient electronics," *Advanced Functional Materials*, **24**(28), pp. 4427–4434, 2014.

[41] C. Dagdeviren, S.-W. Hwang, Y. Su, S. Kim, H. Cheng, O. Gur, R. Haney, F. G. Omenetto, Y. Huang, and J. A. Rogers, "Transient, biocompatible electronics and energy harvesters based on ZnO," *Small*, **9**(20), pp. 3398–3404, 2013.

[42] S.-W. Hwang, S.-K. Kang, X. Huang, M. A. Brenckle, F. G. Omenetto, and J. A. Rogers, "Materials for programmed, functional transformation in transient electronic systems," *Advanced Materials*, **27**(1), pp. 47–52, 2015.

[43] S.-W. Hwang, X. Huang, J.-H. Seo, J.-K. Song, S. Kim, S. Hage-Ali, H.-J. Chung, H. Tao, F. G. Omenetto, Z. Ma, and J. A. Rogers, "Materials for bioresorbable radio frequency electronics," *Advanced Materials*, **25**(26), pp. 3526–3531, 2013.

[44] S.-W. Hwang, J.-K. Song, X. Huang, H. Cheng, S.-K. Kang, B. H. Kim, J.-H. Kim, S. Yu, Y. Huang, and J. A. Rogers, "High-performance biodegradable/transient electronics on biodegradable polymers," *Advanced Materials*, **26**(23), pp. 3905–3911, 2014.

[45] D. Son, J. Lee, D. J. Lee, R. Ghaffari, S. Yun, S. J. Kim, J. E. Lee, H. R. Cho, S. Yoon, S. Yang, S. Lee, S. Qiao, D. Ling, S. Shin, J.-K. Song, J. Kim, T. Kim, H. Lee, J. Kim, M. Soh, N. Lee, C. S. Hwang, S. Nam, N. Lu, T. Hyeon, S. H. Choi, and D.-H. Kim, "Bioresorbable electronic stent integrated with therapeutic nanoparticles for endovascular diseases," *ACS Nano*, **9**(6), pp. 5937–5946, 2015.

[46] C. H. Lee, H. Kim, D. V. Harburg, G. Park, Y. Ma, T. Pan, J. S. Kim, N. Y. Lee, B. H. Kim, K.-I. Jang, S.-K. Kang, Y. Huang, J. Kim, K.-M. Lee, C. Leal, and J. A. Rogers, "Biological lipid membranes for on-demand, wireless drug delivery from thin, bioresorbable electronic implants," *NPG Asia Mater*, **7**, p. e227, 2015.

[47] C. f. D. Control and Prevention, "National hospital discharge survey: 2010," Atlanta (GA): CDC [online]. Available from URL: http://www.cdc.gov/nchs/nhds. htm. [Accessed 2009 Nov. 9.] 2014.

[48] A. R. Salkind and K. C. Rao, "Antiobiotic prophylaxis to prevent surgical site infections," *Am Fam Physician*, **83**(5), pp. 585–590, 2011.

[49] B. H. Robinson, "E-waste: An assessment of global production and environmental impacts," *Science of The Total Environment*, **408**(2), pp. 183-191, 2009.

[50] L. Luther, *Managing Electronic Waste: Issues with Exporting E-Waste*: DIANE Publishing Company, 2010.

[51] E. Spalvins, B. Dubey, and T. Townsend, "Impact of electronic waste disposal on lead concentrations in landfill leachate," *Environmental Science & Technology*, **42**(19), pp. 7452-7458, 2008.

[52] B. R. Babu, A. K. Parande, and C. A. Basha, "Electrical and electronic waste: a global environmental problem," *Waste Management & Research*, **25**(4), pp. 307-318, 2007.

[53] L. S. Morf, J. Tremp, R. Gloor, Y. Huber, M. Stengele, and M. Zennegg, "Brominated flame retardants in waste electrical and electronic equipment: substance flows in a recycling plant," *Environmental Science & Technology*, **39**(22), pp. 8691-8699, 2005.

[54] A. Leung, Z. W. Cai, and M. H. Wong, "Environmental contamination from electronic waste recycling at Guiyu, southeast China," *Journal of Material Cycles and Waste Management*, **8**(1), pp. 21-33, 2006.

[55] R. Widmer, H. Oswald-Krapf, D. Sinha-Khetriwal, M. Schnellmann, and H. Böni, "Global perspectives on e-waste," *Environmental Impact Assessment Review*, **25**(5), pp. 436-458, 2005.

[56] X. Chi, M. Streicher-Porte, M. Y. Wang, and M. A. Reuter, "Informal electronic waste recycling: a sector review with special focus on China," *Waste Management*, **31**(4), pp. 731-742, 2011.

[57] B. K. Reck and T. E. Graedel, "Challenges in metal recycling," *Science*, **337**(6095), pp. 690-695, 2012.

[58] E. Underwood, *Trace Elements in Human and Animal Nutrition 4e*: Elsevier, 2012.

[59] S. Demirel, M. Tuzen, S. Saracoglu, and M. Soylak, "Evaluation of various digestion procedures for trace element contents of some food materials," *Journal of Hazardous Materials*, **152**(3), pp. 1020-1026, 2008.

[60] F. H. Nielsen, "Essential and toxic trace elements in human health and disease," *Current Topics in Nutrition and Disease*, **18**, pp. 277-292, 2008.

[61] N. T. Kirkland, "Magnesium biomaterials: past, present and future," *Corrosion Engineering, Science and Technology*, **47**(5), pp. 322-328, 2012.

[62] A. Hartwig, "Role of magnesium in genomic stability," *Mutation Research/Fundamental and Molecular Mechanisms of Mutagenesis*, **475**(1–2), pp. 113–121, 2001.

[63] C. J. Damien and J. R. Parsons, "Bone graft and bone graft substitutes: A review of current technology and applications," *Journal of Applied Biomaterials*, **2**(3), pp. 187–208, 1991.

[64] M. Bohner, "Resorbable biomaterials as bone graft substitutes," *Materials Today*, **13**(1–2), pp. 24–30, 2010.

[65] F. Witte, "The history of biodegradable magnesium implants: a review," *Acta Biomaterialia*, **6**(5), pp. 1680–1692, 2010.

[66] A. Chaya, S. Yoshizawa, K. Verdelis, N. Myers, B. J. Costello, D.-T. Chou, S. Pal, S. Maiti, P. N. Kumta, and C. Sfeir, "In vivo study of magnesium plate and screw degradation and bone fracture healing," *Acta Biomaterialia*, **18**, pp. 262–269, 2015.

[67] R. Erbel, C. Di Mario, J. Bartunek, J. Bonnier, B. de Bruyne, F. R. Eberli, P. Erne, M. Haude, B. Heublein, M. Horrigan, C. Ilsley, D. Böse, J. Koolen, T. F. Lüscher, N. Weissman, and R. Waksman, "Temporary scaffolding of coronary arteries with bioabsorbable magnesium stents: a prospective, non-randomised multicentre trial," *The Lancet*, **369**(9576), pp. 1869–1875, 2007.

[68] T. L. P. Slottow, R. Pakala, T. Okabe, D. Hellinga, R. J. Lovec, F. O. Tio, A. B. Bui, and R. Waksman, "Optical coherence tomography and intravascular ultrasound imaging of bioabsorbable magnesium stent degradation in porcine coronary arteries," *Cardiovascular Revascularization Medicine*, **9**(4), pp. 248–254, 2008.

[69] C. Di Mario, H. Griffiths, O. Goktekin, N. Peeters, J. Verbist, M. Bosiers, K. Deloose, B. Heublein, R. Rohde, and V. Kasese, "Drug-eluting bioabsorbable magnesium stent," *Journal of Interventional Cardiology*, **17**(6), pp. 391–395, 2004.

[70] H. Tao, S.-W. Hwang, B. Marelli, B. An, J. E. Moreau, M. Yang, M. A. Brenckle, S. Kim, D. L. Kaplan, and J. A. Rogers, "Silk-based resorbable electronic devices for remotely controlled therapy and in vivo infection abatement," *Proceedings of the National Academy of Sciences*, **111**(49), pp. 17385–17389, 2014.

[71] G. Makar and J. Kruger, "*Corrosion of magnesium*," *International Materials Reviews*, 2013.

[72] M. Razavi, M. H. Fathi, O. Savabi, D. Vashaee, and L. Tayebi, "Biodegradation, bioactivity and in vivo biocompatibility

analysis of plasma electrolytic oxidized (PEO) biodegradable Mg implants," *Physical Science International Journal*, **4**(5), p. 708, 2014.

[73] K. Wei Guo, "A review of magnesium/magnesium alloys corrosion," *Recent Patents on Corrosion Science*, **1**(1), pp. 72–90, 2011.

[74] G. Song and S. Song, "A possible biodegradable magnesium implant material," *Advanced Engineering Materials*, **9**(4), pp. 298–302, 2007.

[75] C. Yuen and W. Ip, "Theoretical risk assessment of magnesium alloys as degradable biomedical implants," *Acta Biomaterialia*, **6**(5), pp. 1808–1812, 2010.

[76] D. Pierson, J. Edick, A. Tauscher, E. Pokorney, P. Bowen, J. Gelbaugh, J. Stinson, H. Getty, C. H. Lee, J. Drelich, and J. Goldman, "A simplified in vivo approach for evaluating the bioabsorbable behavior of candidate stent materials," *Journal of Biomedical Materials Research Part B: Applied Biomaterials*, **100B**(1), pp. 58–67, 2012.

[77] J. Gray-Munro and M. Strong, "The mechanism of deposition of calcium phosphate coatings from solution onto magnesium alloy AZ31," *Journal of Biomedical Materials Research Part A*, **90**(2), pp. 339–350, 2009.

[78] Y. Zhang, G. Zhang, and M. Wei, "Controlling the biodegradation rate of magnesium using biomimetic apatite coating," *Journal of Biomedical Materials Research Part B: Applied Biomaterials*, **89**(2), pp. 408–414, 2009.

[79] H. M. Wong, K. W. Yeung, K. O. Lam, V. Tam, P. K. Chu, K. D. Luk, and K. M. Cheung, "A biodegradable polymer-based coating to control the performance of magnesium alloy orthopaedic implants," *Biomaterials*, **31**(8), pp. 2084–2096, 2010.

[80] M. Li, Y. Cheng, Y. Zheng, X. Zhang, T. Xi, and S. Wei, "Surface characteristics and corrosion behaviour of WE43 magnesium alloy coated by SiC film," *Applied Surface Science*, **258**(7), pp. 3074–3081, 2012.

[81] J. Hu, Q. Li, X. Zhong, and W. Kang, "Novel anti-corrosion silicon dioxide coating prepared by sol–gel method for AZ91D magnesium alloy," *Progress in Organic Coatings*, **63**(1), pp. 13–17, 2008.

[82] Y. Song, D. Shan, and E. Han, "Electrodeposition of hydroxyapatite coating on AZ91D magnesium alloy for biomaterial application," *Materials Letters*, **62**(17), pp. 3276–3279, 2008.

[83] J. Gray and B. Luan, "Protective coatings on magnesium and its alloys—a critical review," *Journal of Alloys and Compounds*, **336**(1), pp. 88–113, 2002.

[84] H. Altun and S. Sen, "The effect of DC magnetron sputtering AlN coatings on the corrosion behaviour of magnesium alloys," *Surface and Coatings Technology*, **197**(2), pp. 193–200, 2005.

[85] G. Song, "Control of biodegradation of biocompatable magnesium alloys," *Corrosion Science*, **49**(4), pp. 1696–1701, 2007.

[86] N. T. Kirkland, N. Birbilis, J. Walker, T. Woodfield, G. J. Dias, and M. P. Staiger, "In-vitro dissolution of magnesium–calcium binary alloys: Clarifying the unique role of calcium additions in bioresorbable magnesium implant alloys," *Journal of Biomedical Materials Research Part B: Applied Biomaterials*, **95B**(1), pp. 91–100, 2010.

[87] S. Zhang, X. Zhang, C. Zhao, J. Li, Y. Song, C. Xie, H. Tao, Y. Zhang, Y. He, Y. Jiang, and Y. Bian, "Research on an Mg–Zn alloy as a degradable biomaterial," *Acta Biomaterialia*, **6**(2), pp. 626–640, 2010.

[88] J. M. Seitz, R. Eifler, J. Stahl, M. Kietzmann, and F. W. Bach, "Characterization of $MgNd_2$ alloy for potential applications in bioresorbable implantable devices," *Acta Biomaterialia*, **8**(10), pp. 3852–3864, 2012.

[89] J. Nie, X. Gao, and S.-M. Zhu, "Enhanced age hardening response and creep resistance of Mg–Gd alloys containing Zn," *Scripta Materialia*, **53**(9), pp. 1049–1053, 2005.

[90] H. S. Brar, M. O. Platt, M. Sarntinoranont, P. I. Martin, and M. V. Manuel, "Magnesium as a biodegradable and bioabsorbable material for medical implants," *JOM*, **61**(9), pp. 31–34, 2009.

[91] J. D. Cao, N. T. Kirkland, K. J. Laws, N. Birbilis, and M. Ferry, "Ca–Mg–Zn bulk metallic glasses as bioresorbable metals," *Acta Biomaterialia*, **8**(6), pp. 2375–2383, 2012.

[92] H. Li, Y. Zheng, and L. Qin, "Progress of biodegradable metals," *Progress in Natural Science: Materials International*, **24**(5), pp. 414–422, 2014.

[93] R. S. MacDonald, "The role of zinc in growth and cell proliferation," *The Journal of Nutrition*, **130**(5), pp. 1500S–1508S, 2000.

[94] M. S. Wold, "Replication protein A: a heterotrimeric, single-stranded DNA-binding protein required for eukaryotic DNA metabolism," *Annual Review of Biochemistry*, **66**(1), pp. 61–92, 1997.

[95] F. Wu and C.-W. Wu, "Zinc in DNA replication and transcription," *Annual Review of Nutrition*, **7**(1), pp. 251–272, 1987.

[96] M. Yamaguchi, "Role of zinc in bone formation and bone resorption," *The Journal of Trace Elements in Experimental Medicine*, **11**(2–3), pp. 119–135, 1998.

[97] J. Brandão-Neto, V. Stefan, B. B. Mendonça, W. Bloise, and A. V. B. Castro, "The essential role of zinc in growth," *Nutrition Research*, **15**(3), pp. 335–358, 1995.

[98] P. K. Bowen, R. J. Guillory Ii, E. R. Shearier, J.-M. Seitz, J. Drelich, M. Bocks, F. Zhao, and J. Goldman, "Metallic zinc exhibits optimal biocompatibility for bioabsorbable endovascular stents," *Materials Science and Engineering: C*, **56**, pp.467–472, 2015.

[99] Y. Yun, Z. Dong, D. Yang, M. J. Schulz, V. N. Shanov, S. Yarmolenko, Z. Xu, P. Kumta, and C. Sfeir, "Biodegradable Mg corrosion and osteoblast cell culture studies," *Materials Science and Engineering: C*, **29**(6), pp. 1814–1821, 2009.

[100] N. Pistofidis, G. Vourlias, S. Konidaris, E. Pavlidou, A. Stergiou, and G. Stergioudis, "The effect of bismuth on the structure of zinc hot-dip galvanized coatings," *Materials Letters*, **61**(4–5), pp. 994–997, 2007.

[101] X. Zhang, S. Lin, X.-Q. Lu, and Z.-l. Chen, "Removal of Pb(II) from water using synthesized kaolin supported nanoscale zero-valent iron," *Chemical Engineering Journal*, **163**(3), pp. 243–248, 2010.

[102] P. K. Bowen, J. Drelich, and J. Goldman, "Zinc exhibits ideal physiological corrosion behavior for bioabsorbable stents," *Advanced Materials*, **25**(18), pp. 2577–2582, 2013.

[103] D. Vojtěch, J. Kubásek, J. Šerák, and P. Novák, "Mechanical and corrosion properties of newly developed biodegradable Zn-based alloys for bone fixation," *Acta Biomaterialia*, **7**(9), pp. 3515–3522, 2011.

[104] K. Törne, M. Larsson, A. Norlin, and J. Weissenrieder, "Degradation of zinc in saline solutions, plasma, and whole blood," *Journal of Biomedical Materials Research Part B: Applied Biomaterials*, **104**(6), pp. 1141–1151, 2016.

[105] B. Hennig, M. Toborek, and C. J. McClain, "Antiatherogenic properties of zinc: implications in endothelial cell metabolism," *Nutrition*, **12**(10), pp. 711–717, 1996.

[106] X. Liu, J. Sun, Y. Yang, Z. Pu, and Y. Zheng, "In vitro investigation of ultra-pure Zn and its mini-tube as potential bioabsorbable stent material," *Materials Letters*, **161**, pp. 53–56, 2015.

[107] L. Zhao, Z. Zhang, Y. Song, S. Liu, Y. Qi, X. Wang, Q. Wang, and C. Cui, "Mechanical properties and in vitro biodegradation of newly

developed porous Zn scaffolds for biomedical applications," *Materials & Design*, **108**, pp. 136–144, 2016.

[108] D. Vojtěch, J. Kubasek, J. Šerák, and P. Novak, "Mechanical and corrosion properties of newly developed biodegradable Zn-based alloys for bone fixation," *Acta Biomaterialia*, **7**(9), pp. 3515–3522, 2011.

[109] A. Bolz and T. Popp, "Implantable, bioresorbable vessel wall support, in particular coronary stent," Google Patents, 2001.

[110] R. Othmán, A. Yahaya, and A. K. Arof, "A zinc–air cell employing a porous zinc electrode fabricated from zinc–graphite-natural biodegradable polymer paste," *Journal of Applied Electrochemistry*, **32**(12), pp. 1347–1353, 2002.

[111] X. Huang, Y. Liu, S.-W. Hwang, S.-K. Kang, D. Patnaik, J. F. Cortes, and J. A. Rogers, "Biodegradable materials for multilayer transient printed circuit boards," *Advanced Materials*, **26**(43), pp. 7371–7377, 2014.

[112] A. Bianco, K. Kostarelos, and M. Prato, "Making carbon nanotubes biocompatible and biodegradable," *Chemical Communications*, **47**(37), pp. 10182–10188, 2011.

[113] I. Jesion, M. Skibniewski, E. Skibniewska, W. Strupiński, L. Szulc-Dąbrowska, A. Krajewska, I. Pasternak, P. Kowalczyk, and R. Pińkowski, "Graphene and carbon nanocompounds: biofunctionalization and applications in tissue engineering," *Biotechnology & Biotechnological Equipment*, **29**(3), pp. 415–422, 2015.

[114] Q. Wang, C. Wang, M. Zhang, M. Jian, and Y. Zhang, "Feeding single-walled carbon nanotubes or graphene to silkworms for reinforced silk fibers," *Nano Letters*, **16**(10), pp. 6695–6700, 2016.

[115] F. Rancan, D. Papakostas, S. Hadam, S. Hackbarth, T. Delair, C. Primard, B. Verrier, W. Sterry, U. Blume-Peytavi, and A. Vogt, "Investigation of polylactic acid (PLA) nanoparticles as drug delivery systems for local dermatotherapy," *Pharmaceutical Research*, **26**(8), pp. 2027–2036, 2009.

[116] K. Oksman, M. Skrifvars, and J. F. Selin, "Natural fibres as reinforcement in polylactic acid (PLA) composites," *Composites Science and Technology*, **63**(9), pp. 1317–1324, 2003.

[117] Y. Cheng, S. Deng, P. Chen, and R. Ruan, "Polylactic acid (PLA) synthesis and modifications: a review," *Frontiers of Chemistry in China*, **4**(3), pp. 259–264, 2009.

[118] S. Shawe, F. Buchanan, E. Harkin-Jones, and D. Farrar, "A study on the rate of degradation of the bioabsorbable polymer polyglycolic acid (PGA)," *Journal of Materials Science*, **41**(15), pp. 4832–4838, 2006.

[119] A. W. T. Shum and A. F. T. Mak, "Morphological and biomechanical characterization of poly(glycolic acid) scaffolds after in vitro degradation," *Polymer Degradation and Stability*, **81**(1), pp. 141–149, 2003.

[120] R. M. Day, A. R. Boccaccini, S. Shurey, J. A. Roether, A. Forbes, L. L. Hench, and S. M. Gabe, "Assessment of polyglycolic acid mesh and bioactive glass for soft-tissue engineering scaffolds," *Biomaterials*, **25**(27), pp. 5857–5866, 2004.

[121] S. Sarkar, G. Y. Lee, J. Y. Wong, and T. A. Desai, "Development and characterization of a porous micro-patterned scaffold for vascular tissue engineering applications," *Biomaterials*, **27**(27), pp. 4775–4782, 2006.

[122] J.-M. Lü, X. Wang, C. Marin-Muller, H. Wang, P. H. Lin, Q. Yao, and C. Chen, "Current advances in research and clinical applications of PLGA-based nanotechnology," *Expert Review of Molecular Diagnostics*, **9**(4), pp. 325–341, 2009.

[123] T. D. Roy, J. L. Simon, J. L. Ricci, E. D. Rekow, V. P. Thompson, and J. R. Parsons, "Performance of degradable composite bone repair products made via three-dimensional fabrication techniques," *Journal of Biomedical Materials Research Part A*, **66A**(2), pp. 283–291, 2003.

[124] S. de Valence, J.-C. Tille, D. Mugnai, W. Mrowczynski, R. Gurny, M. Möller, and B. H. Walpoth, "Long term performance of poly-caprolactone vascular grafts in a rat abdominal aorta replacement model," *Biomaterials*, **33**(1), pp. 38–47, 2012.

[125] K. H. Lee, H. Y. Kim, M. S. Khil, Y. M. Ra, and D. R. Lee, "Characterization of nano-structured poly(ε-caprolactone) non-woven mats via electrospinning," *Polymer*, **44**(4), pp. 1287–1294, 2003.

[126] J. S. Chawla and M. M. Amiji, "Biodegradable poly(ε-caprolactone) nanoparticles for tumor-targeted delivery of tamoxifen," *International Journal of Pharmaceutics*, **249**(1-2), pp. 127–138, 2002.

[127] K. Zhao, Y. Deng, J. Chun Chen, and G.-Q. Chen, "Polyhydroxyalkanoate (PHA) scaffolds with good mechanical

properties and biocompatibility," *Biomaterials*, **24**(6), pp. 1041–1045, 2003.

[128] M. Zinn, B. Witholt, and T. Egli, "Occurrence, synthesis and medical application of bacterial polyhydroxyalkanoate," *Advanced Drug Delivery Reviews*, **53**(1), pp. 5–21, 2001.

[129] E. I. Shishatskaya, T. G. Volova, A. P. Puzyr, O. A. Mogilnaya, and S. N. Efremov, "Tissue response to the implantation of biodegradable polyhydroxyalkanoate sutures," *Journal of Materials Science: Materials in Medicine*, **15**(6), pp. 719–728, 2004.

[130] L. Tan, X. Yu, P. Wan, and K. Yang, "Biodegradable materials for bone repairs: A review," *Journal of Materials Science & Technology*, **29**(6), pp. 503–513, 2013.

[131] A. J. R. Lasprilla, G. A. R. Martinez, B. H. Lunelli, A. L. Jardini, and R. M. Filho, "Poly-lactic acid synthesis for application in biomedical devices – A review," *Biotechnology Advances*, **30**(1), pp. 321–328, 2012.

[132] Z. G. Tang, R. A. Black, J. M. Curran, J. A. Hunt, N. P. Rhodes, and D. F. Williams, "Surface properties and biocompatibility of solvent-cast poly[ε-caprolactone] films," *Biomaterials*, **25**(19), pp. 4741–4748, 2004.

[133] C. Hwang, Y. Park, J. Park, K. Lee, K. Sun, A. Khademhosseini, and S. H. Lee, "Controlled cellular orientation on PLGA microfibers with defined diameters," *Biomedical Microdevices*, **11**(4), pp. 739–746, 2009.

[134] Y. Zhang, H. Ouyang, C. T. Lim, S. Ramakrishna, and Z. M. Huang, "Electrospinning of gelatin fibers and gelatin/PCL composite fibrous scaffolds," *Journal of Biomedical Materials Research Part B: Applied Biomaterials*, **72**(1), pp. 156–165, 2005.

[135] J.-W. Rhim, A. K. Mohanty, S. P. Singh, and P. K. W. Ng, "Effect of the processing methods on the performance of polylactide films: Thermocompression versus solvent casting," *Journal of Applied Polymer Science*, **101**(6), pp. 3736–3742, 2006.

[136] A. M. Harris and E. C. Lee, "Improving mechanical performance of injection molded PLA by controlling crystallinity," *Journal of Applied Polymer Science*, **107**(4), pp. 2246–2255, 2008.

[137] V. Maquet and R. Jerome, "Design of macroporous biodegradable polymer scaffolds for cell transplantation," in *Materials Science Forum*, 1997, pp. 15–42.

[138] L. D. Harris, B.-S. Kim, and D. J. Mooney, "Open pore biodegradable matrices formed with gas foaming," *Journal of Biomedical Materials Research*, **42**(3), pp. 396–402, 1998.

[139] T. K. Kim, J. J. Yoon, D. S. Lee, and T. G. Park, "Gas foamed open porous biodegradable polymeric microspheres," *Biomaterials*, **27**(2), pp. 152–159, 2006.

[140] F. Danhier, E. Ansorena, J. M. Silva, R. Coco, A. Le Breton, and V. Préat, "PLGA-based nanoparticles: An overview of biomedical applications," *Journal of Controlled Release*, **161**(2), pp. 505–522, 2012.

[141] J. Hu, M. P. Prabhakaran, L. Tian, X. Ding, and S. Ramakrishna, "Drug-loaded emulsion electrospun nanofibers: characterization, drug release and in vitro biocompatibility," *RSC Advances*, **5**(121), pp. 100256–100267, 2015.

[142] L. Peponi, I. Navarro-Baena, A. Sonseca, E. Gimenez, A. Marcos-Fernandez, and J. M. Kenny, "Synthesis and characterization of PCL–PLLA polyurethane with shape memory behavior," *European Polymer Journal*, **49**(4), pp. 893–903, 2013.

[143] X. Yu, L. Wang, M. Huang, T. Gong, W. Li, Y. Cao, D. Ji, P. Wang, J. Wang, and S. Zhou, "A shape memory stent of poly(ε-caprolactone-co-dl-lactide) copolymer for potential treatment of esophageal stenosis," *Journal of Materials Science: Materials in Medicine*, **23**(2), pp. 581–589, 2012.

[144] W. Wang, P. Ping, X. Chen, and X. Jing, "Biodegradable polyurethane based on random copolymer of L-lactide and ε-caprolactone and its shape-memory property," *Journal of Applied Polymer Science*, **104**(6), pp. 4182–4187, 2007.

[145] D. Cohn and A. Hotovely Salomon, "Designing biodegradable multiblock PCL/PLA thermoplastic elastomers," *Biomaterials*, **26**(15), pp. 2297–2305, 2005.

[146] S. H. Choi and T. G. Park, "Synthesis and characterization of elastic PLGA/PCL/PLGA tri-block copolymers," *Journal of Biomaterials Science, Polymer Edition*, **13**(10), pp. 1163–1173, 2002.

[147] E. I. Shishatskaya, T. G. Volova, S. A. Gordeev, and A. P. Puzyr, "Degradation of P(3HB) and P(3HB-co-3HV) in biological media," *Journal of Biomaterials Science, Polymer Edition*, **16**(5), pp. 643–657, 2005.

[148] S. P. Valappil, S. K. Misra, A. R. Boccaccini, and I. Roy, "Biomedical applications of polyhydroxyalkanoates, an overview of animal

testing and in vivo responses," *Expert Review of Medical Devices,* **3**
(6), pp. 853–868, 2006.

[149] S. Philip, T. Keshavarz, and I. Roy, "Polyhydroxyalkanoates: biodegradable polymers with a range of applications," *Journal of
Chemical Technology & Biotechnology,* **82**(3), pp. 233–247, 2007.

[150] C. Hinüber, K. Chwalek, F. J. Pan-Montojo, M. Nitschke, R. Vogel,
H. Brünig, G. Heinrich, and C. Werner, "Hierarchically structured
nerve guidance channels based on poly-3-hydroxybutyrate
enhance oriented axonal outgrowth," *Acta Biomaterialia,* **10**(5),
pp. 2086–2095, 2014.

[151] R. Nigmatullin, P. Thomas, B. Lukasiewicz, H. Puthussery, and I.
Roy, "Polyhydroxyalkanoates, a family of natural polymers, and
their applications in drug delivery," *Journal of Chemical
Technology & Biotechnology,* **90**(7), pp. 1209–1221, 2015.

[152] L. Francis, D. Meng, J. Knowles, T. Keshavarz, A. R. Boccaccini, and
I. Roy, "Controlled delivery of gentamicin using poly (3-hydroxy-
butyrate) microspheres," *International Journal of Molecular
Sciences,* **12**(7), pp. 4294–4314, 2011.

[153] J. An, K. Wang, S. Chen, M. Kong, Y. Teng, L. Wang, C. Song, D.
Kong, and S. Wang, "Biodegradability, cellular compatibility and
cell infiltration of poly (3-hydroxybutyrate-co-4-hydroxybutyrate)
in comparison with poly (ε-caprolactone) and poly (lactide-co-
glycolide)," *Journal of Bioactive and Compatible Polymers:
Biomedical Applications,* **30**(2), pp. 209–221, 2015.

[154] B. D. Ulery, L. S. Nair, and C. T. Laurencin, "Biomedical applications of biodegradable polymers," *Journal of Polymer Science Part
B: Polymer Physics,* **49**(12), pp. 832–864, 2011.

[155] M. Kellomäki, H. Niiranen, K. Puumanen, N. Ashammakhi,
T. Waris, and P. Törmälä, "Bioabsorbable scaffolds for guided
bone regeneration and generation," *Biomaterials,* **21**(24), pp.
2495–2505, 2000.

[156] M. H. Sheridan, L. D. Shea, M. C. Peters, and D. J. Mooney,
"Bioabsorbable polymer scaffolds for tissue engineering capable
of sustained growth factor delivery," *Journal of Controlled Release,*
**64**(1–3), pp. 91–102, 2000.

[157] M. van der Elst, C. P. A. T. Klein, J. M. de Blieck-Hogervorst,
P. Patka, and H. J. T. M. Haarman, "Bone tissue response to biodegradable polymers used for intra medullary fracture fixation:

A long-term in vivo study in sheep femora," *Biomaterials*, **20**(2), pp. 121–128, 1999.

[158] S. Vainionpää, J. Kilpikari, J. Laiho, P. Helevirta, P. Rokkanen, and P. Törmälä, "Strength and strength retention vitro, of absorbable, self-reinforced polyglycolide (PGA) rods for fracture fixation," *Biomaterials*, **8**(1), pp. 46–48, 1987.

[159] B. Rai, S. H. Teoh, D. W. Hutmacher, T. Cao, and K. H. Ho, "Novel PCL-based honeycomb scaffolds as drug delivery systems for rhBMP-2," *Biomaterials*, **26**(17), pp. 3739–3748, 2005.

[160] E. Grube, S. Sonoda, F. Ikeno, Y. Honda, S. Kar, C. Chan, U. Gerckens, A. J. Lansky, and P. J. Fitzgerald, "Six-and twelve-month results from first human experience using everolimus-eluting stents with bioabsorbable polymer," *Circulation*, **109**(18), pp. 2168–2171, 2004.

[161] P. Erne, M. Schier, and T. J. Resink, "The road to bioabsorbable stents: Reaching clinical reality?," *CardioVascular and Interventional Radiology*, **29**(1), pp. 11–16, 2006.

[162] C. J. Bettinger and Z. Bao, "Organic thin-film transistors fabricated on resorbable biomaterial substrates," *Advanced Materials*, **22**(5), pp. 651–655, 2010.

[163] A. Campana, T. Cramer, D. T. Simon, M. Berggren, and F. Biscarini, "Electrocardiographic recording with conformable organic electrochemical transistor fabricated on resorbable bioscaffold," *Advanced Materials*, **26**(23), pp. 3874–3878, 2014.

[164] K. J. Yu, D. Kuzum, S.-W. Hwang, B. H. Kim, H. Juul, N. H. Kim, S. M. Won, K. Chiang, M. Trumpis, A. G. Richardson, H. Cheng, H. Fang, M. Thompson, H. Bink, D. Talos, K. J. Seo, H. N. Lee, S.-K. Kang, J.-H. Kim, J. Y. Lee, Y. Huang, F. E. Jensen, M. A. Dichter, T. H. Lucas, J. Viventi, B. Litt, and J. A. Rogers, "Bioresorbable silicon electronics for transient spatiotemporal mapping of electrical activity from the cerebral cortex," *Nat Mater*, **15**(7), pp. 782–791, 2016.

[165] L. Yin, X. Huang, H. Xu, Y. Zhang, J. Lam, J. Cheng, and J. A. Rogers, "Materials, designs, and operational characteristics for fully biodegradable primary batteries," *Advanced Materials*, **26**(23), pp. 3879–3884, 2014.

[166] S.-K. Kang, R. K. J. Murphy, S.-W. Hwang, S. M. Lee, D. V. Harburg, N. A. Krueger, J. Shin, P. Gamble, H. Cheng, S. Yu, Z. Liu, J. G. McCall, M. Stephen, H. Ying, J. Kim, G. Park, R. C. Webb, C. H. Lee,

S. Chung, D. S. Wie, A. D. Gujar, B. Vemulapalli, A. H. Kim, K.-M. Lee, J. Cheng, Y. Huang, S. H. Lee, P. V. Braun, W. Z. Ray, and J. A. Rogers, "Bioresorbable silicon electronic sensors for the brain," *Nature*, **530**(7588), pp. 71–76, 2016.

[167] W. E. Hennink and C. F. van Nostrum, "Novel crosslinking methods to design hydrogels," *Advanced Drug Delivery Reviews*, **64**, Supplement, pp. 223–236, 2012.

[168] J. Zhu, "Bioactive modification of poly(ethylene glycol) hydrogels for tissue engineering," *Biomaterials*, **31**(17), pp. 4639–4656, 2010.

[169] A. Revzin, R. J. Russell, V. K. Yadavalli, W.-G. Koh, C. Deister, D. D. Hile, M. B. Mellott, and M. V. Pishko, "Fabrication of poly(ethylene glycol) hydrogel microstructures using photolithography," *Langmuir*, **17**(18), pp. 5440–5447, 2001.

[170] W.-G. Koh, A. Revzin, and M. V. Pishko, "Poly(ethylene glycol) hydrogel microstructures encapsulating living cells," *Langmuir*, **18**(7), pp. 2459–2462, 2002.

[171] H. Otsuka, Y. Nagasaki, and K. Kataoka, "Self-assembly of poly(ethylene glycol)-based block copolymers for biomedical applications," *Current Opinion in Colloid & Interface Science*, **6**(1), pp. 3–10, 2001.

[172] S. N. S. Alconcel, A. S. Baas, and H. D. Maynard, "FDA-approved poly(ethylene glycol)-protein conjugate drugs," *Polymer Chemistry*, **2**(7), pp. 1442–1448, 2011.

[173] M. B. Mellott, K. Searcy, and M. V. Pishko, "Release of protein from highly cross-linked hydrogels of poly(ethylene glycol) diacrylate fabricated by UV polymerization," *Biomaterials*, **22**(9), pp. 929–941, 2001.

[174] A. Revzin, R. G. Tompkins, and M. Toner, "Surface engineering with poly(ethylene glycol) photolithography to create high-density cell arrays on glass," *Langmuir*, **19**(23), pp. 9855–9862, 2003.

[175] K. T. Nguyen and J. L. West, "Photopolymerizable hydrogels for tissue engineering applications," *Biomaterials*, **23**(22), pp. 4307–4314, 2002.

[176] B. K. Mann, A. S. Gobin, A. T. Tsai, R. H. Schmedlen, and J. L. West, "Smooth muscle cell growth in photopolymerized hydrogels with cell adhesive and proteolytically degradable domains: synthetic ECM analogs for tissue engineering," *Biomaterials*, **22**(22), pp. 3045–3051, 2001.

[177] A. K. Gaharwar, C. P. Rivera, C.-J. Wu, and G. Schmidt, "Transparent, elastomeric and tough hydrogels from poly(ethylene

glycol) and silicate nanoparticles," *Acta Biomaterialia*, **7**(12), pp. 4139–4148, 2011.

[178] T. Fujiwara, T. Mukose, T. Yamaoka, H. Yamane, S. Sakurai, and Y. Kimura, "Novel thermo-responsive formation of a hydrogel by stereo-complexation between PLLA-PEG-PLLA and PDLA-PEG-PDLA block copolymers," *Macromolecular Bioscience*, **1**(5), pp. 204–208, 2001.

[179] K. Nagahama, K. Fujiura, S. Enami, T. Ouchi, and Y. Ohya, "Irreversible temperature-responsive formation of high-strength hydrogel from an enantiomeric mixture of starburst triblock copolymers consisting of 8-arm PEG and PLLA or PDLA," *Journal of Polymer Science Part A: Polymer Chemistry*, **46**(18), pp. 6317–6332, 2008.

[180] C. Gong, S. Shi, L. Wu, M. Gou, Q. Yin, Q. Guo, P. Dong, F. Zhang, F. Luo, and X. Zhao, "Biodegradable in situ gel-forming controlled drug delivery system based on thermosensitive PCL–PEG–PCL hydrogel. Part 2: Sol–gel–sol transition and drug delivery behavior," *Acta Biomaterialia*, **5**(9), pp. 3358–3370, 2009.

[181] C. B. Liu, C. Y. Gong, M. J. Huang, J. W. Wang, Y. F. Pan, Y. D. Zhang, G. Z. Li, M. L. Gou, K. Wang, and M. J. Tu, "Thermoreversible gel-sol behavior of biodegradable PCL–PEG–PCL triblock copolymer in aqueous solutions," *Journal of Biomedical Materials Research Part B: Applied Biomaterials*, **84**(1), pp. 165–175, 2008.

[182] M. Qiao, D. Chen, X. Ma, and Y. Liu, "Injectable biodegradable temperature-responsive PLGA–PEG–PLGA copolymers: Synthesis and effect of copolymer composition on the drug release from the copolymer-based hydrogels," *International Journal of Pharmaceutics*, **294**(1–2), pp. 103–112, 2005.

[183] S. Choi, M. Baudys, and S. W. Kim, "Control of blood glucose by novel GLP-1 delivery using biodegradable triblock copolymer of PLGA-PEG-PLGA in type 2 diabetic rats," *Pharmaceutical Research*, **21**(5), pp. 827–831, 2004.

[184] A. Douglas, N. Muralidharan, R. Carter, K. Share, and C. L. Pint, "Ultrafast triggered transient energy storage by atomic layer deposition into porous silicon for integrated transient electronics," *Nanoscale*, **8**(14), pp. 7384–7390, 2016.

[185] K.-H. Kim, L. Jeong, H.-N. Park, S.-Y. Shin, W.-H. Park, S.-C. Lee, T.-I. Kim, Y.-J. Park, Y.-J. Seol, Y.-M. Lee, Y. Ku, I.-C. Rhyu, S.-B. Han, and C.-P. Chung, "Biological efficacy of silk fibroin nanofiber

membranes for guided bone regeneration," *Journal of Biotechnology*, **120**(3), pp. 327–339, 2005.

[186] E. Wenk, H. P. Merkle, and L. Meinel, "Silk fibroin as a vehicle for drug delivery applications," *Journal of Controlled Release*, **150**(2), pp. 128–141, 2011.

[187] C. H. F. Hämmerle and N. P. Lang, "Single stage surgery combining transmucosal implant placement with guided bone regeneration and bioresorbable materials," *Clinical Oral Implants Research*, **12**(1), pp. 9–18, 2001.

[188] S. A. Sell, M. J. McClure, K. Garg, P. S. Wolfe, and G. L. Bowlin, "Electrospinning of collagen/biopolymers for regenerative medicine and cardiovascular tissue engineering," *Advanced Drug Delivery Reviews*, **61**(12), pp. 1007–1019, 2009.

[189] A. Kuijpers, P. Van Wachem, M. Van Luyn, J. Plantinga, G. Engbers, J. Krijgsveld, S. Zaat, J. Dankert, and J. Feijen, "In vivo compatibility and degradation of crosslinked gelatin gels incorporated in knitted Dacron," *Journal of Biomedical Materials Research*, **51**(1), pp. 136–145, 2000.

[190] A. Duconseille, T. Astruc, N. Quintana, F. Meersman, and V. Sante-Lhoutellier, "Gelatin structure and composition linked to hard capsule dissolution: a review," *Food Hydrocolloids*, **43**, pp. 360–376, 2015.

[191] S. Partridge and H. Davis, "The chemistry of connective tissues. 3. Composition of the soluble proteins derived from elastin," *Biochemical Journal*, **61**(1), p. 21, 1955.

[192] C. N. Grover, R. E. Cameron, and S. M. Best, "Investigating the morphological, mechanical and degradation properties of scaffolds comprising collagen, gelatin and elastin for use in soft tissue engineering," *Journal of the Mechanical Behavior of Biomedical Materials*, **10**, pp. 62–74, 2012.

[193] J. W. Chang, C. G. Wang, C. Y. Huang, T. Tzung-Da, T. F. Guo, and T. C. Wen, "Chicken albumen dielectrics in organic field-effect transistors," *Advanced Materials*, **23**(35), pp. 4077–81, 2011.

[194] M. Li, M. J. Mondrinos, M. R. Gandhi, F. K. Ko, A. S. Weiss, and P. I. Lelkes, "Electrospun protein fibers as matrices for tissue engineering," *Biomaterials*, **26**(30), pp. 5999–6008, 2005.

[195] W. Qiu, Y. Huang, W. Teng, C. M. Cohn, J. Cappello, and X. Wu, "Complete recombinant silk-elastinlike protein-based tissue scaffold," *Biomacromolecules*, **11**(12), pp. 3219–3227, 2010.

[196] I.-S. Yeo, J.-E. Oh, L. Jeong, T. S. Lee, S. J. Lee, W. H. Park, and B.-M. Min, "Collagen-based biomimetic nanofibrous scaffolds: Preparation and characterization of collagen/silk fibroin bicomponent nanofibrous structures," *Biomacromolecules*, **9**(4), pp. 1106–1116, 2008.

[197] M. Zilberman, N. D. Schwade, and R. C. Eberhart, "Protein-loaded bioresorbable fibers and expandable stents: Mechanical properties and protein release," *Journal of Biomedical Materials Research Part B: Applied Biomaterials*, **69B**(1), pp. 1–10, 2004.

[198] D. Asai, D. Xu, W. Liu, F. Garcia Quiroz, D. J. Callahan, M. R. Zalutsky, S. L. Craig, and A. Chilkoti, "Protein polymer hydrogels by in situ, rapid and reversible self-gelation," *Biomaterials*, **33**(21), pp. 5451–5458, 2012.

[199] B. Kundu, R. Rajkhowa, S. C. Kundu, and X. Wang, "Silk fibroin biomaterials for tissue regenerations," *Advanced Drug Delivery Reviews*, **65**(4), pp. 457–470, 2013.

[200] Y. Gui-Bo, Z. You-Zhu, W. Shu-Dong, S. De-Bing, D. Zhi-Hui, and F. Wei-Guo, "Study of the electrospun PLA/silk fibroin-gelatin composite nanofibrous scaffold for tissue engineering," *Journal of Biomedical Materials Research Part A*, **93**(1), pp. 158–163, 2010.

[201] H.-Y. Cheung, K.-T. Lau, X.-M. Tao, and D. Hui, "A potential material for tissue engineering: Silkworm silk/PLA biocomposite," *Composites Part B: Engineering*, **39**(6), pp. 1026–1033, 2008.

[202] M. Li, M. J. Mondrinos, X. Chen, M. R. Gandhi, F. K. Ko, and P. I. Lelkes, "Co-electrospun poly (lactide-co-glycolide), gelatin, and elastin blends for tissue engineering scaffolds," *Journal of Biomedical Materials Research Part A*, **79**(4), pp. 963–973, 2006.

[203] Z. Meng, Y. Wang, C. Ma, W. Zheng, L. Li, and Y. Zheng, "Electrospinning of PLGA/gelatin randomly-oriented and aligned nanofibers as potential scaffold in tissue engineering," *Materials Science and Engineering: C*, **30**(8), pp. 1204–1210, 2010.

[204] L. Li, H. Li, Y. Qian, X. Li, G. K. Singh, L. Zhong, W. Liu, Y. Lv, K. Cai, and L. Yang, "Electrospun poly (ε-caprolactone)/silk fibroin core-sheath nanofibers and their potential applications in tissue engineering and drug release," *International Journal of Biological Macromolecules*, **49**(2), pp. 223–232, 2011.

[205] D.-B. Jeon, J.-Y. Bak, and S.-M. Yoon, "Oxide thin-film transistors fabricated using biodegradable gate dielectric layer of chicken

albumen," *Japanese Journal of Applied Physics*, **52**(12 R), p. 128002, 2013.

[206] R. Capelli, J. J. Amsden, G. Generali, S. Toffanin, V. Benfenati, M. Muccini, D. Kaplan, F. Omenetto, and R. Zamboni, "Integration of silk protein in organic and light-emitting transistors," *Organic Electronics*, **12**(7), pp. 1146–1151, 2011.

[207] W. S. Lour, W. C. Liu, J. H. Tsai, and L. W. Laih, "High-performance camel-gate field effect transistor using high-medium-low doped structure," *Applied Physics Letters*, **67**(18), pp. 2636–2638, 1995.

[208] L.-K. Mao, J.-C. Hwang, T.-H. Chang, C.-Y. Hsieh, L.-S. Tsai, Y.-L. Chueh, S. S. Hsu, P.-C. Lyu, and T.-J. Liu, "Pentacene organic thin-film transistors with solution-based gelatin dielectric," *Organic Electronics*, **14**(4), pp. 1170–1176, 2013.

[209] W.-H. Zhang, B.-J. Jiang, and P. Yang, "Proteins as functional interlayer in organic field-effect transistor," *Chinese Chemical Letters*.

[210] H. Im, X.-J. Huang, B. Gu, and Y.-K. Choi, "A dielectric-modulated field-effect transistor for biosensing," *Nature Nanotechnology*, **2**(7), pp. 430–434, 2007.

[211] P. Hu, A. Fasoli, J. Park, Y. Choi, P. Estrela, S. L. Maeng, W. I. Milne, and A. C. Ferrari, "Self-assembled nanotube field-effect transistors for label-free protein biosensors," *Journal of Applied Physics*, **104**(7), p. 074310, 2008.

[212] T. Minamiki, T. Minami, P. Koutnik, P. Anzenbacher Jr, and S. Tokito, "Antibody- and label-free phosphoprotein sensor device based on an organic transistor," *Analytical Chemistry*, **88**(2), pp. 1092–1095, 2016.

[213] C. C. Cid, J. Riu, A. Maroto, and F. X. Rius, "Carbon nanotube field effect transistors for the fast and selective detection of human immunoglobulin G," *Analyst*, **133**(8), pp. 1005–1008, 2008.

[214] K.-Y. Park, Y.-S. Sohn, C.-K. Kim, H.-S. Kim, Y.-S. Bae, and S.-Y. Choi, "Development of FET-type albumin sensor for diagnosing nephritis," *Biosensors and Bioelectronics*, **23**(12), pp. 1904–1907, 2008.

[215] J. Chen, K. Vongsanga, X. Wang, and N. Byrne, "What happens during natural protein fibre dissolution in ionic liquids," *Materials*, **7**(9), pp. 6158–6168, 2014.

[216] D. N. Rockwood, R. C. Preda, T. Yucel, X. Wang, M. L. Lovett, and D. L. Kaplan, "Materials fabrication from *Bombyx mori* silk fibroin," *Nat. Protocols*, **6**(10), pp. 1612–1631, 2011.

[217] S. Keten, Z. Xu, B. Ihle, and M. J. Buehler, "Nanoconfinement controls stiffness, strength and mechanical toughness of [beta]-sheet crystals in silk," *Nat Mater*, **9**(4), pp. 359–367, 2010.

[218] T. Lefèvre, M.-E. Rousseau, and M. Pézolet, "Protein secondary structure and orientation in silk as revealed by Raman spectro-microscopy," *Biophysical Journal*, **92**(8), pp. 2885–2895, 2007.

[219] X. Hu, K. Shmelev, L. Sun, E.-S. Gil, S.-H. Park, P. Cebe, and D. L. Kaplan, "Regulation of silk material structure by temperature-controlled water vapor annealing," *Biomacromolecules*, **12**(5), pp. 1686–1696, 2011.

[220] M. Li, M. Ogiso, and N. Minoura, "Enzymatic degradation behavior of porous silk fibroin sheets," *Biomaterials*, **24**(2), pp. 357–365, 2003.

[221] T. Arai, G. Freddi, R. Innocenti, and M. Tsukada, "Biodegradation of *Bombyx mori* silk fibroin fibers and films," *Journal of Applied Polymer Science*, **91**(4), pp. 2383–2390, 2004.

[222] K. Chen, Y. Umeda, and K. Hirabayashi, "Enzymatic hydrolysis of silk fibroin," *The Journal of Sericultural Science of Japan*, **65**(2), pp. 131–133, 1996.

[223] E. M. Pritchard and D. L. Kaplan, "Silk fibroin biomaterials for controlled release drug delivery," *Expert Opinion on Drug Delivery*, **8**(6), pp. 797–811, 2011.

[224] X. Wang, T. Yucel, Q. Lu, X. Hu, and D. L. Kaplan, "Silk nanospheres and microspheres from silk/pva blend films for drug delivery," *Biomaterials*, **31**(6), pp. 1025–1035, 2010.

[225] A. S. Lammel, X. Hu, S.-H. Park, D. L. Kaplan, and T. R. Scheibel, "Controlling silk fibroin particle features for drug delivery," *Biomaterials*, **31**(16), pp. 4583–4591, 2010.

[226] S. Enomoto, M. Sumi, K. Kajimoto, Y. Nakazawa, R. Takahashi, C. Takabayashi, T. Asakura, and M. Sata, "Long-term patency of small-diameter vascular graft made from fibroin, a silk-based biodegradable material," *Journal of Vascular Surgery*, **51**(1), pp. 155–164, 2010.

[227] Y. Nakazawa, M. Sato, R. Takahashi, D. Aytemiz, C. Takabayashi, T. Tamura, S. Enomoto, M. Sata, and T. Asakura, "Development of small-diameter vascular grafts based on silk fibroin fibers from *Bombyx mori* for vascular regeneration," *Journal of Biomaterials Science, Polymer Edition*, **22**(1–3), pp. 195–206, 2011.

[228] K. Gruchenberg, A. Ignatius, B. Friemert, F. von Lübken, N. Skaer, K. Gellynck, O. Kessler, and L. Dürselen, "In vivo performance of a

novel silk fibroin scaffold for partial meniscal replacement in a sheep model," *Knee Surgery, Sports Traumatology, Arthroscopy*, **23**(8), pp. 2218–2229, 2015.

[229] Y. Wang, U.-J. Kim, D. J. Blasioli, H.-J. Kim, and D. L. Kaplan, "In vitro cartilage tissue engineering with 3D porous aqueous-derived silk scaffolds and mesenchymal stem cells," *Biomaterials*, **26**(34), pp. 7082–7094, 2005.

[230] G. H. Altman, F. Diaz, C. Jakuba, T. Calabro, R. L. Horan, J. Chen, H. Lu, J. Richmond, and D. L. Kaplan, "Silk-based biomaterials," *Biomaterials*, **24**(3), pp. 401–416, 2003.

[231] M. S. Mannoor, H. Tao, J. D. Clayton, A. Sengupta, D. L. Kaplan, R. R. Naik, N. Verma, F. G. Omenetto, and M. C. McAlpine, "Graphene-based wireless bacteria detection on tooth enamel," *Nature Communications*, **3**, p. 763, 2012.

[232] F. G. Omenetto and D. L. Kaplan, "A new route for silk," *Nature Photonics*, **2**(11), pp. 641–643, 2008.

[233] S. T. Parker, P. Domachuk, J. Amsden, J. Bressner, J. A. Lewis, D. L. Kaplan, and F. G. Omenetto, "Biocompatible silk printed optical waveguides," *Advanced Materials*, **21**(23), pp. 2411–2415, 2009.

[234] H. Tao, J. J. Amsden, A. C. Strikwerda, K. Fan, D. L. Kaplan, X. Zhang, R. D. Averitt, and F. G. Omenetto, "Metamaterial silk composites at terahertz frequencies," *Advanced Materials*, **22**(32), pp. 3527–3531, 2010.

[235] G. A. Digenis, T. B. Gold, and V. P. Shah, "Cross-linking of gelatin capsules and its relevance to their in vitro–in vivo performance," *Journal of Pharmaceutical Sciences*, **83**(7), pp. 915–921, 1994.

[236] D. L. Casey, R. M. Beihn, G. A. Digenis, and M. B. Shambhu, "Method for monitoring hard gelatin capsule disintegration times in humans using external scintigraphy," *Journal of Pharmaceutical Sciences*, **65**(9), pp. 1412–1413, 1976.

[237] K. B. Djagny, Z. Wang, and S. Xu, "Gelatin: a valuable protein for food and pharmaceutical industries: review," *Critical Reviews in Food Science and Nutrition*, **41**(6), pp. 481–492, 2001.

[238] J. E. Botzolakis and L. L. Augsburger, "Disintegrating agents in hard gelatin capsules. Part II: Swelling efficiency," *Drug Development and Industrial Pharmacy*, **14**(9), pp. 1235–1248, 1988.

[239] X. Lou and T. V. Chirila, "Swelling behavior and mechanical properties of chemically cross-linked gelatin gels for biomedical use," *Journal of Biomaterials Applications*, **14**(2), pp. 184–191, 1999.

[240] K. Y. Lee, J. Shim, and H. G. Lee, "Mechanical properties of gellan and gelatin composite films," *Carbohydrate Polymers*, **56**(2), pp. 251–254, 2004.

[241] F. Hom, S. Veresh, and J. Miskel, "Soft gelatin capsules I: Factors affecting capsule shell dissolution rate," *Journal of Pharmaceutical Sciences*, **62**(6), pp. 1001–1006, 1973.

[242] E. Negrete-Abascal, V. R. Tenorio, J. J. Serrano, C. Garcia, and M. de la Garza, "Secreted proteases from Actinobacillus pleuropneumoniae serotype 1 degrade porcine gelatin, hemoglobin and immunoglobulin A," *Canadian Journal of Veterinary Research*, **58**(2), p. 83, 1994.

[243] Y. Tabata and Y. Ikada, "Protein release from gelatin matrices," *Advanced Drug Delivery Reviews*, **31**(3), pp. 287–301, 1998.

[244] M. Irimia-Vladu, P. A. Troshin, M. Reisinger, G. Schwabegger, M. Ullah, R. Schwoediauer, A. Mumyatov, M. Bodea, J. W. Fergus, and V. F. Razumov, "Environmentally sustainable organic field effect transistors," *Organic Electronics*, **11**(12), pp. 1974–1990, 2010.

[245] C. Uhlig, M. Rapp, B. Hartmann, H. Hierlemann, H. Planck, and K.-K. Dittel, "Suprathel® – An innovative, resorbable skin substitute for the treatment of burn victims," *Burns*, **33**(2), pp. 221–229, 2007.

[246] P. Zahedi, I. Rezaeian, S.-O. Ranaei-Siadat, S.-H. Jafari, and P. Supaphol, "A review on wound dressings with an emphasis on electrospun nanofibrous polymeric bandages," *Polymers for Advanced Technologies*, **21**(2), pp. 77–95, 2010.

[247] R. A. Allen, W. Wu, M. Yao, D. Dutta, X. Duan, T. N. Bachman, H. C. Champion, D. B. Stolz, A. M. Robertson, and K. Kim, "Nerve regeneration and elastin formation within poly (glycerol sebacate)-based synthetic arterial grafts one-year post-implantation in a rat model," *Biomaterials*, **35**(1), pp. 165–173, 2014.

[248] J. Yang, D. Motlagh, J. B. Allen, A. R. Webb, M. R. Kibbe, O. Aalami, M. Kapadia, T. J. Carroll, and G. A. Ameer, "Modulating expanded polytetrafluoroethylene vascular graft host response via citric acid-based biodegradable elastomers," *Advanced Materials*, **18**(12), pp. 1493–1498, 2006.

[249] R. Rai, M. Tallawi, A. Grigore, and A. R. Boccaccini, "Synthesis, properties and biomedical applications of poly(glycerol sebacate) (PGS): A review," *Progress in Polymer Science*, **37**(8), pp. 1051–1078, 2012.

[250] Y. Kang, J. Yang, S. Khan, L. Anissian, and G. A. Ameer, "A new biodegradable polyester elastomer for cartilage tissue engineering," *Journal of Biomedical Materials Research Part A*, **77A**(2), pp. 331–339, 2006.

[251] R. Rai, M. Tallawi, N. Barbani, C. Frati, D. Madeddu, S. Cavalli, G. Graiani, F. Quaini, J. A. Roether, and D. W. Schubert, "Biomimetic poly (glycerol sebacate)(PGS) membranes for cardiac patch application," *Materials Science and Engineering: C*, **33**(7), pp. 3677–3687, 2013.

[252] M. P. Prabhakaran, A. S. Nair, D. Kai, and S. Ramakrishna, "Electrospun composite scaffolds containing poly (octanediol-co-citrate) for cardiac tissue engineering," *Biopolymers*, **97**(7), pp. 529–538, 2012.

[253] P. M. Crapo, J. Gao, and Y. Wang, "Seamless tubular poly (glycerol sebacate) scaffolds: High-yield fabrication and potential applications," *Journal of Biomedical Materials Research Part A*, **86**(2), pp. 354–363, 2008.

[254] K.-W. Lee, D. B. Stolz, and Y. Wang, "Substantial expression of mature elastin in arterial constructs," *Proceedings of the National Academy of Sciences*, **108**(7), pp. 2705–2710, 2011.

[255] S.-L. Chia, K. Gorna, S. Gogolewski, and M. Alini, "Biodegradable elastomeric polyurethane membranes as chondrocyte carriers for cartilage repair," *Tissue Engineering*, **12**(7), pp. 1945–1953, 2006.

[256] S. Grad, L. Kupcsik, K. Gorna, S. Gogolewski, and M. Alini, "The use of biodegradable polyurethane scaffolds for cartilage tissue engineering: potential and limitations," *Biomaterials*, **24**(28), pp. 5163–5171, 2003.

[257] M. Borkenhagen, R. Stoll, P. Neuenschwander, U. Suter, and P. Aebischer, "In vivo performance of a new biodegradable polyester urethane system used as a nerve guidance channel," *Biomaterials*, **19**(23), pp. 2155–2165, 1998.

[258] B. Amsden, "Curable, biodegradable elastomers: emerging biomaterials for drug delivery and tissue engineering," *Soft Matter*, **3**(11), pp. 1335–1348, 2007.

[259] K. Gorna and S. Gogolewski, "Biodegradable porous polyurethane scaffolds for tissue repair and regeneration," *Journal of Biomedical Materials Research Part A*, **79A**(1), pp. 128–138, 2006.

[260] V. Kanyanta and A. Ivankovic, "Mechanical characterisation of polyurethane elastomer for biomedical applications," *Journal of*

*the Mechanical Behavior of Biomedical Materials*, **3**(1), pp. 51–62, 2010.

[261] E. Bat, B. H. M. Kothman, G. A. Higuera, C. A. van Blitterswijk, J. Feijen, and D. W. Grijpma, "Ultraviolet light crosslinking of poly (trimethylene carbonate) for elastomeric tissue engineering scaffolds," *Biomaterials*, **31**(33), pp. 8696–8705, 2010.

[262] B. L. Dargaville, C. d. Vaquette, H. Peng, F. Rasoul, Y. Q. Chau, J. J. Cooper-White, J. H. Campbell, and A. K. Whittaker, "Cross-linked poly (trimethylene carbonate-co-L-lactide) as a biodegradable, elastomeric scaffold for vascular engineering applications," *Biomacromolecules*, **12**(11), pp. 3856–3869, 2011.

[263] C. J. Bettinger, B. Orrick, A. Misra, R. Langer, and J. T. Borenstein, "Microfabrication of poly (glycerol–sebacate) for contact guidance applications," *Biomaterials*, **27**(12), pp. 2558–2565, 2006.

[264] D. P. Martin and S. F. Williams, "Medical applications of poly-4-hydroxybutyrate: a strong flexible absorbable biomaterial," *Biochemical Engineering Journal*, **16**(2), pp. 97–105, 2003.

[265] S. F. Williams, S. Rizk, and D. P. Martin, "Poly-4-hydroxybutyrate (P4HB): a new generation of resorbable medical devices for tissue repair and regeneration," *Biomedizinische Technik/Biomedical Engineering*, **58**(5), pp. 1–14, 2013.

[266] J. Yang, A. R. Webb, S. J. Pickerill, G. Hageman, and G. A. Ameer, "Synthesis and evaluation of poly (diol citrate) biodegradable elastomers," *Biomaterials*, **27**(9), pp. 1889–1898, 2006.

[267] A. Patel, A. K. Gaharwar, G. Iviglia, H. Zhang, S. Mukundan, S. M. Mihaila, D. Demarchi, and A. Khademhosseini, "Highly elastomeric poly(glycerol sebacate)-co-poly(ethylene glycol) amphiphilic block copolymers," *Biomaterials*, **34**(16), pp. 3970–3983, 2013.

[268] M. C. Serrano, E. J. Chung, and G. Ameer, "Advances and applications of biodegradable elastomers in regenerative medicine," *Advanced Functional Materials*, **20**(2), pp. 192–208, 2010.

[269] Y. Wang, G. A. Ameer, B. J. Sheppard, and R. Langer, "A tough biodegradable elastomer," *Nat Biotech*, **20**(6), pp. 602–606, 2002.

[270] S. Sant, C. M. Hwang, S.-H. Lee, and A. Khademhosseini, "Hybrid PGS–PCL microfibrous scaffolds with improved mechanical and biological properties," *Journal of Tissue Engineering and Regenerative Medicine*, **5**(4), pp. 283–291, 2011.

[271] S.-L. Liang, X.-Y. Yang, X.-Y. Fang, W. D. Cook, G. A. Thouas, and Q.-Z. Chen, "In vitro enzymatic degradation of poly (glycerol sebacate)-based materials," *Biomaterials*, **32**(33), pp. 8486–8496, 2011.

[272] C. M. Boutry, A. Nguyen, Q. O. Lawal, A. Chortos, and Z. Bao, "Fully biodegradable pressure sensor, viscoelastic behavior of PGS dielectric elastomer upon degradation," in *SENSORS, 2015 IEEE*, 2015, pp. 1–4.

[273] J. Yang, A. R. Webb, and G. A. Ameer, "Novel citric acid-based biodegradable elastomers for tissue engineering," *Advanced Materials*, **16**(6), pp. 511–516, 2004.

[274] J. Yang, A. R. Webb, S. J. Pickerill, G. Hageman, and G. A. Ameer, "Synthesis and evaluation of poly(diol citrate) biodegradable elastomers," *Biomaterials*, **27**(9), pp. 1889–1898, 2006.

[275] S.-W. Hwang, C. H. Lee, H. Cheng, J.-W. Jeong, S.-K. Kang, J.-H. Kim, J. Shin, J. Yang, Z. Liu, G. A. Ameer, Y. Huang, and J. A. Rogers, "Biodegradable elastomers and silicon nanomembranes/nanoribbons for stretchable, transient electronics, and biosensors," *Nano Letters*, **15**(5), pp. 2801–2808, 2015.

[276] R. B. Reed, D. A. Ladner, C. P. Higgins, P. Westerhoff, and J. F. Ranville, "Solubility of nano-zinc oxide in environmentally and biologically important matrices," *Environmental Toxicology and Chemistry*, **31**(1), pp. 93–99, 2012.

[277] T. Xia, M. Kovochich, M. Liong, L. Mädler, B. Gilbert, H. Shi, J. I. Yeh, J. I. Zink, and A. E. Nel, "Comparison of the mechanism of toxicity of zinc oxide and cerium oxide nanoparticles based on dissolution and oxidative stress properties," *ACS Nano*, **2**(10), pp. 2121–2134, 2008.

[278] K. R. Raghupathi, R. T. Koodali, and A. C. Manna, "Size-dependent bacterial growth inhibition and mechanism of antibacterial activity of zinc oxide nanoparticles," *Langmuir*, **27**(7), pp. 4020–4028, 2011.

[279] A. Janotti and C. G. Van de Walle, "Fundamentals of zinc oxide as a semiconductor," *Reports on Progress in Physics*, **72**(12), p. 126501, 2009.

[280] S. Roy and S. Basu, "Improved zinc oxide film for gas sensor applications," *Bulletin of Materials Science*, **25**(6), pp. 513–515, 2002.

[281] Z. L. Wang and J. Song, "Piezoelectric nanogenerators based on zinc oxide nanowire arrays," *Science*, **312**(5771), pp. 242–246, 2006.

[282] J. A. Mejias, A. J. Berry, K. Refson, and D. G. Fraser, "The kinetics and mechanism of MgO dissolution," *Chemical Physics Letters*, **314**(5-6), pp. 558–563, 1999.

[283] A. Fedoročková and P. Raschman, "Effects of pH and acid anions on the dissolution kinetics of MgO," *Chemical Engineering Journal*, **143**(1-3), pp. 265–272, 2008.

[284] J. Fontanella, C. Andeen, and D. Schuele, "Low-frequency dielectric constants of $\alpha$-quartz, sapphire, MgF2, and MgO," *Journal of Applied Physics*, **45**(7), pp. 2852–2854, 1974.

[285] L. Yan, C. M. Lopez, R. P. Shrestha, E. A. Irene, A. A. Suvorova, and M. Saunders, "Magnesium oxide as a candidate high-κ gate dielectric," *Applied Physics Letters*, **88**(14), p. 142901, 2006.

[286] A. Posadas, F. J. Walker, C. H. Ahn, T. L. Goodrich, Z. Cai, and K. S. Ziemer, "Epitaxial MgO as an alternative gate dielectric for SiC transistor applications," *Applied Physics Letters*, **92**(23), p. 233511, 2008.

[287] Y. Irokawa, Y. Nakano, M. Ishiko, T. Kachi, J. Kim, F. Ren, B. P. Gila, A. H. Onstine, C. R. Abernathy, S. J. Pearton, C.-C. Pan, G.-T. Chen, and J.-I. Chyi, "MgO/p-GaN enhancement mode metal-oxide semi-conductor field-effect transistors," *Applied Physics Letters*, **84**(15), pp. 2919–2921, 2004.

[288] H. Jagannathan, V. Narayanan, and S. Brown, "Engineering high dielectric constant materials for band-edge CMOS applications," *ECS Transactions*, **16**(5), pp. 19–26, 2008.

[289] R. Villota, J. G. Hawkes, and H. Cochrane, "Food applications and the toxicological and nutritional implications of amorphous silicon dioxide," *C R C Critical Reviews in Food Science and Nutrition*, **23**(4), pp. 289–321, 1986.

[290] M. G. Shahram, W. T. Benjamin, E. U. Ronald, O. Carina, K. Thomas, B. Mike, M. Ralph, and C. J. Kirkpatrick, "Collagen-embedded hydro-xylapatite–beta-tricalcium phosphate–silicon dioxide bone substi-tute granules assist rapid vascularization and promote cell growth," *Biomedical Materials*, **5**(2), p. 025004, 2010.

[291] P. V. Giannoudis, H. Dinopoulos, and E. Tsiridis, "Bone substitutes: An update," *Injury*, **36**(3, Supplement), pp. S20–S27, 2005.

[292] G. Li, S. Feng, and D. Zhou, "Magnetic bioactive glass ceramic in the system $CaO-P_2O_5-SiO_2-MgO-CaF_2-MnO_2-Fe_2O_3$ for hyperthermia treatment of bone tumor," *Journal of Materials Science: Materials in Medicine*, **22**(10), pp. 2197–2206, 2011.

[293] T. W. Wang, H. C. Wu, W. R. Wang, F. H. Lin, P. J. Lou, M. J. Shieh, and T. H. Young, "The development of magnetic degradable DP-bioglass for hyperthermia cancer therapy," *Journal of Biomedical Materials Research Part A*, **83**(3), pp. 828-837, 2007.

[294] F. J. Martin, K. Melnik, T. West, J. Shapiro, M. Cohen, A. A. Boiarski, and M. Ferrari, "Acute toxicity of intravenously administered microfabricated silicon dioxide drug delivery particles in mice," *Drugs in R & D*, **6**(2), pp. 71-81, 2005.

[295] Y. Li, Y.-Z. Liu, T. Long, X.-B. Yu, T. T. Tang, K.-R. Dai, B. Tian, Y.-P. Guo, and Z.-A. Zhu, "Mesoporous bioactive glass as a drug delivery system: fabrication, bactericidal properties and biocompatibility," *Journal of Materials Science: Materials in Medicine*, **24**(8), pp. 1951-1961, 2013.

[296] E. J. Anglin, L. Cheng, W. R. Freeman, and M. J. Sailor, "Porous silicon in drug delivery devices and materials," *Advanced Drug Delivery Reviews*, **60**(11), pp. 1266-1277, 2008.

[297] J. D. Birchall and J. S. Chappell, "The chemistry of aluminum and silicon in relation to Alzheimer's disease," *Clin Chem*, **34**(2), pp. 265-267, 1988.

[298] K. S. Finnie, D. J. Waller, F. L. Perret, A. M. Krause-Heuer, H. Q. Lin, J. V. Hanna, and C. J. Barbé, "Biodegradability of sol–gel silica microparticles for drug delivery," *Journal of Sol–Gel Science and Technology*, **49**(1), pp. 12-18, 2009.

[299] S. K. Kang, S. W. Hwang, H. Cheng, S. Yu, B. H. Kim, J. H. Kim, Y. Huang, and J. A. Rogers, "Dissolution behaviors and applications of silicon oxides and nitrides in transient electronics," *Advanced Functional Materials*, **24**(28), pp. 4427-4434, 2014.

[300] B. S. Bal and M. N. Rahaman, "Orthopedic applications of silicon nitride ceramics," *Acta Biomaterialia*, **8**(8), pp. 2889-2898, 2012.

[301] J. Olofsson, T. M. Grehk, T. Berlind, C. Persson, S. Jacobson, and H. Engqvist, "Evaluation of silicon nitride as a wear resistant and resorbable alternative for total hip joint replacement," *Biomatter*, **2**(2), pp. 94-102, 2012.

[302] C. C. Guedes e Silva, B. König Jr, M. J. Carbonari, M. Yoshimoto, S. Allegrini Jr, and J. C. Bressiani, "Bone growth around silicon nitride implants – An evaluation by scanning electron microscopy," *Materials Characterization*, **59**(9), pp. 1339-1341, 2008.

[303] C. C. Guedes e Silva, O. Z. Higa, and J. C. Bressiani, "Cytotoxic evaluation of silicon nitride-based ceramics," *Materials Science and Engineering: C*, **24**(5), pp.643–646, 2004.

[304] Y. Yee Chia, L. Qiang, L. Wen Chin, K. Tsu-Jae, H. Chenming, W. Xiewen, G. Xin, and T. P. Ma, "Direct tunneling gate leakage current in transistors with ultrathin silicon nitride gate dielectric," *IEEE Electron Device Letters*, **21**(11), pp. 540–542, 2000.

[305] F. M. Li, A. Nathan, Y. Wu, and B. S. Ong, "Organic thin-film transistor integration using silicon nitride gate dielectric," *Applied Physics Letters*, **90**(13), p. 133514, 2007.

[306] M. She, H. Takeuchi, and T.-J. King, "Silicon-nitride as a tunnel dielectric for improved SONOS-type flash memory," *IEEE Electron Device Letters*, **24**(5), pp. 309–311, 2003.

[307] H. Aozasa, I. Fujiwara, and Y. Komatsu, "Analysis of carrier traps in Si3N4 in oxide/nitride/oxide for metal/oxide/nitride/oxide/silicon nonvolatile memory," *Japanese Journal of Applied Physics*, **38**(3R), p. 1441, 1999.

[308] M. A. Whitehead, D. Fan, P. Mukherjee, G. R. Akkaraju, L. T. Canham, and J. L. Coffer, "High-Porosity poly($\varepsilon$-caprolactone)/ mesoporous silicon scaffolds: calcium phosphate deposition and biological response to bone precursor cells," *Tissue Engineering Part A*, **14**(1), pp. 195–206, 2008.

[309] D. Liang, J. Wang, and Y. Wang, "Difference in dissolution between germanium and zinc during the oxidative pressure leaching of sphalerite," *Hydrometallurgy*, **95**(1–2), pp. 5–7, 2009.

[310] W. W. Harvey and H. C. Gatos, "The reaction of germanium with aqueous solutions: I. Dissolution kinetics in water containing dissolved oxygen," *Journal of the Electrochemical Society*, **105**(11), pp. 654–660, 1958.

[311] S.-K. Kang, G. Park, K. Kim, S.-W. Hwang, H. Cheng, J. Shin, S. Chung, M. Kim, L. Yin, J. C. Lee, K.-M. Lee, and J. A. Rogers, "Dissolution chemistry and biocompatibility of silicon- and germanium-based semiconductors for transient electronics," *ACS Applied Materials & Interfaces*, **7**(17), pp. 9297–9305, 2015.

[312] J. Versieck and J. T. McCall, "Trace elements in human body fluids and tissues," *CRC Critical Reviews in Clinical Laboratory Sciences*, **22**(2), pp. 97–184, 1985.

[313] J. A. Pennington, "Silicon in foods and diets," *Food Addit Contam*, **8**(1), pp. 97–118, 1991.

*evaporation–condensation-mediated laser printing and sintering of Zn nanoparticles," Advanced Materials,* 2017.

[334] Y. J. Kim, S.-E. Chun, J. Whitacre, and C. J. Bettinger, "Self-deployable current sources fabricated from edible materials," *Journal of Materials Chemistry B,* **1**(31), pp. 3781–3788, 2013.

[335] X. Jia, Y. Yang, C. Wang, C. Zhao, R. Vijayaraghavan, D. R. MacFarlane, M. Forsyth, and G. G. Wallace, "Biocompatible ionic liquid–biopolymer electrolyte-enabled thin and compact magnesium–air batteries," *ACS Applied Materials & Interfaces,* **6**(23), pp. 21110–21117, 2014.

[336] M. Tsang, A. Armutlulu, A. W. Martinez, S. A. B. Allen, and M. G. Allen, "Biodegradable magnesium/iron batteries with polycaprolactone encapsulation: A microfabricated power source for transient implantable devices," *Microsystems & Nanoengineering,* **1**, p. 15024, 2015.

[337] R. K. Pal, A. A. Farghaly, C. Wang, M. M. Collinson, S. C. Kundu, and V. K. Yadavalli, "Conducting polymer–silk biocomposites for flexible and biodegradable electrochemical sensors," *Biosensors and Bioelectronics,* **81**, pp. 294–302, 2016.

[338] M. Luo, A. W. Martinez, C. Song, F. Herrault, and M. G. Allen, "A microfabricated wireless RF pressure sensor made completely of biodegradable materials," *Journal of Microelectromechanical Systems,* **23**(1), pp. 4–13, 2014.

[339] H. Tao, M. A. Brenckle, M. Yang, J. Zhang, M. Liu, S. M. Siebert, R. D. Averitt, M. S. Mannoor, M. C. McAlpine, J. A. Rogers, D. L. Kaplan, and F. G. Omenetto, "Silk-based conformal, adhesive, edible food sensors," *Advanced Materials,* **24**(8), pp. 1067–1072, 2012.

[340] S.-W. Hwang, D.-H. Kim, H. Tao, T.-i. Kim, S. Kim, K. J. Yu, B. Panilaitis, J.-W. Jeong, J.-K. Song, F. G. Omenetto, and J. A. Rogers, "Materials and fabrication processes for transient and bioresorbable high-performance electronics," *Advanced Functional Materials,* **23**(33), pp. 4087–4093, 2013.

[341] S.-K. Kang, S.-W. Hwang, S. Yu, J.-H. Seo, E. A. Corbin, J. Shin, D. S. Wie, R. Bashir, Z. Ma, and J. A. Rogers, "Biodegradable thin metal foils and spin-on glass materials for transient electronics," *Advanced Functional Materials,* **25**(12), pp. 1789–1797, 2015.

[342] J. Guo, J. Liu, B. Yang, G. Zhan, L. Tang, H. Tian, X. Kang, H. Peng, X. Chen, and C. Yang, "biodegradable junctionless transistors with extremely simple structure," *IEEE Electron Device Letters*, **36**(9), pp. 908–910, 2015.

[343] R. Capelli, J. J. Amsden, G. Generali, S. Toffanin, V. Benfenati, M. Muccini, D. L. Kaplan, F. G. Omenetto, and R. Zamboni, "Integration of silk protein in organic and light-emitting transistors," *Organic Electronics*, **12**(7), pp. 1146–1151, 2011.

[344] M. Irimia-Vladu, N. S. Sariciftci, and S. Bauer, "Exotic materials for bio-organic electronics," *Journal of Materials Chemistry*, **21**(5), pp. 1350–1361, 2011.

[345] H. Sirringhaus, T. Kawase, R. H. Friend, T. Shimoda, M. Inbasekaran, W. Wu, and E. P. Woo, "High-resolution inkjet printing of all-polymer transistor circuits," *Science*, **290**(5499), pp. 2123–2126, 2000.

[346] B. J. de Gans, P. C. Duineveld, and U. S. Schubert, "Inkjet printing of polymers: state of the art and future developments," *Advanced Materials*, **16**(3), pp. 203–213, 2004.

[347] E. Tekin, P. J. Smith, and U. S. Schubert, "Inkjet printing as a deposition and patterning tool for polymers and inorganic particles," *Soft Matter*, **4**(4), pp. 703–713, 2008.

[348] H. K. Seung, P. Heng, P. G. Costas, K. L. Christine, M. J. F. Jean, and P. Dimos, "All-inkjet-printed flexible electronics fabrication on a polymer substrate by low-temperature high-resolution selective laser sintering of metal nanoparticles," *Nanotechnology*, **18**(34), p. 345202, 2007.

[349] M. S. Rill, C. Plet, M. Thiel, I. Staude, G. von Freymann, S. Linden, and M. Wegener, "Photonic metamaterials by direct laser writing and silver chemical vapour deposition," *Nat Mater*, **7**(7), pp. 543–546, 2008.

[350] A. I. Kuznetsov, A. B. Evlyukhin, M. R. Gonçalves, C. Reinhardt, A. Koroleva, M. L. Arnedillo, R. Kiyan, O. Marti, and B. N. Chichkov, "Laser fabrication of large-scale nanoparticle arrays for sensing applications," *ACS Nano*, **5**(6), pp. 4843–4849, 2011.

[351] Y. Galagan, E. W. C. Coenen, R. Abbel, T. J. van Lammeren, S. Sabik, M. Barink, E. R. Meinders, R. Andriessen, and P. W. M. Blom, "Photonic sintering of inkjet printed current collecting grids for organic solar cell applications," *Organic Electronics*, **14**(1), pp. 38–46, 2013.

[352] M. Hosel and F. C. Krebs, "Large-scale roll-to-roll photonic sinter-ing of flexo printed silver nanoparticle electrodes," *Journal of Materials Chemistry*, **22**(31), pp. 15683–15688, 2012.

[353] W.-S. Han, J.-M. Hong, H.-S. Kim, and Y.-W. Song, "Multi-pulsed white light sintering of printed Cu nanoinks," *Nanotechnology*, **22**(39), p. 395705, 2011.

Printed in the United States
By Bookmasters